Astronomers' Universe

For other titles published in this series, go to
http://www.springer.com/series/6960

Olmes Bisi

Visible and Invisible

The Wonders of Light Phenomena

 Springer

Olmes Bisi
Science and Methods for Engineering
University of Modena and Reggio Emilia
Modena, Reggio Emilia, Italy

Originally published in Italian by SironiEditore, 2011

ISSN 1614-659X ISSN 2197-6651 (electronic)
ISBN 978-3-319-09824-1 ISBN 978-3-319-09825-8 (eBook)
DOI 10.1007/978-3-319-09825-8
Springer Cham Heidelberg New York Dordrecht London

Library of Congress Control Number: 2014950897

© Springer International Publishing Switzerland 2015
This work is subject to copyright. All rights are reserved by the Publisher, whether the whole or part of the material is concerned, specifically the rights of translation, reprinting, reuse of illustrations, recitation, broadcasting, reproduction on microfilms or in any other physical way, and transmission or information storage and retrieval, electronic adaptation, computer software, or by similar or dissimilar methodology now known or hereafter developed. Exempted from this legal reservation are brief excerpts in connection with reviews or scholarly analysis or material supplied specifically for the purpose of being entered and executed on a computer system, for exclusive use by the purchaser of the work. Duplication of this publication or parts thereof is permitted only under the provisions of the Copyright Law of the Publisher's location, in its current version, and permission for use must always be obtained from Springer. Permissions for use may be obtained through RightsLink at the Copyright Clearance Center. Violations are liable to prosecution under the respective Copyright Law.
The use of general descriptive names, registered names, trademarks, service marks, etc. in this publication does not imply, even in the absence of a specific statement, that such names are exempt from the relevant protective laws and regulations and therefore free for general use.
While the advice and information in this book are believed to be true and accurate at the date of publication, neither the authors nor the editors nor the publisher can accept any legal responsibility for any errors or omissions that may be made. The publisher makes no warranty, express or implied, with respect to the material contained herein.

Printed on acid-free paper

Springer is part of Springer Science+Business Media (www.springer.com)

To Matteo

Preface

Light is an essential presence for life. It has a strong symbolic value and is in many ways a very fascinating entity. Since ancient times, humanity has wondered about its nature, developing models and theories stretching between myth and science.

The bright sky was the first scientific laboratory of humanity, the only one for millennia. The discovery of electromagnetic waves broadened the scope of interest with the notion of invisible light.

This book stems from a range of different experiences, from my teaching work in physics university courses to that of researcher in the field of optoelectronics. However, my primary inspiration was triggered by the discovery of an innovative approach to education, known as the Reggio Approach. In this context, I participated in the Reggio Children projects for the creation of two new environments and contexts, called Ateliers: Ray of Light and From Wave to Wave, the first at the 'Loris Malaguzzi' International Center of Reggio Emilia, the second, dedicated to water and energy, based at the hydroelectric power plant at Ligonchio on the Tuscan-Emilian Apennine Ridge.

The challenge faced by the ateliers is to involve every individual in creative investigation, not just the few who constitute the 'elite,' those who visit science museums or read science books.

How can you interest everyone? How can you create experiences that stimulate curiosity? By providing rewarding activities and encouraging curiosity without harnessing reasoning and also by promoting free-form investigations that on one hand develop the cognitive sphere and on the other give satisfaction through creative work. Such activities are not static nor always the same but are inserted into contexts where any result can be different from what has been seen before, responding to both new and old questions.

For these reasons, *Visible and Invisible* does not offer a single way of reading but can be used in different ways, even drawing on individual chapters, shifting from one topic to another on the basis of the interest of the moment. The book also offers several possible insights, with different incentives for further investigation and consideration prompted by the statement CONSIDER THIS. Readers are thereby encouraged to seek out their own explanations (the 'answers' are not given in the text). Author's views can be found in the website www.visibleandinvisible.eu.

The basic hypothesis of the atelier and also of this book is that the investigation of nature can be both creative and enjoyable. Barbara McClintock (1902–1992), the American biologist and winner of the Nobel Prize for Medicine in 1983 for her discovery of transposons, described her life as a scientist in this way: "I was just so interested in what I was doing I could hardly wait to get up in the morning and get at it. One of my friends, a geneticist, said I was a child, because only children can't wait to get up in the morning to get at what they want to do." In fact, children and scientists are not so different: they both enjoy discovering the world.

Let us look at what the 'Ray of Light' atelier is in concrete terms. This facility is offered to children, young people, students, teachers and visitors of all kinds as a place of research, experimentation and immersion in an environment where light is investigated in its various perceptive, emotional and rational forms. Didactically, it stands out by virtue of the fact that the discovery paths within the atelier are not predetermined (Figs. P1, P2 and P3).

Visitors are not required to learn the laws of nature but get closer to them through their own sensitivity, their own personal approach. The role of educators and 'atelieristas' is limited to stimulating creativity, without influencing the formation of reasoning and theories.

The atelier is set out as follows:

- Welcome areas, where the project is shared with schools and visitors.
- Perception areas for the exploration and immersion in the theme of light (Fig. P4).
- Areas of in-depth conceptual and research analysis and interaction called illuminators.

Preface ix

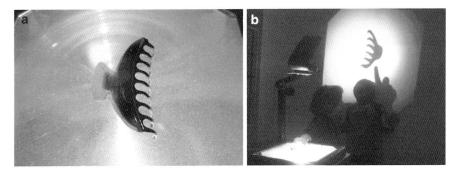

FIGURE P1 (**a, b**) This shows some phases of the experience of Marco and Lucia, two children of 4 and 5 years of age attending the 'Gulliver' preschool in Reggio Emilia. Their teacher was Lucia Levrini Lucia, the 'atelierista' was Anna Orlandini and the educator was Maddalena Tedeschi. The experiment started with a question: what happens to an object (in this case a comb) when it is placed on the overhead projector? How can we come to terms with the two worlds—that of the real object and that of its projected image?

FIGURE P2 After various tests and repositioning of the object, Marco (M) and Lucia (L) make their first observations

- Areas for the discussion and sharing of the experiences and documents created in the research areas.

    M: *"Its shadow becomes different."*
    L: *"It leans in a different direction."*

x   Preface

FIGURE P3 The conversation continued

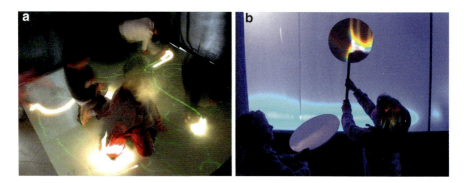

FIGURE P4 (**a**, **b**) Some moments from the visit of a group of children to the Ray of Light atelier

Both capture the change in the orientation of the comb, due to reflection.

M: "I think there must be an electronic system, that travels along all this way to get high up and then it comes to the wall."
L: "It's the system that transports the picture of the object."

Marco offers an initial explanation: the electronic system itself, inscrutable and powerful; while Lucia points out the difference between objects and their image: "the figure of the objects."

The atelier was presented during the Science Festival in Genoa in 2005, 2006 and 2008 and is situated in the laboratory area of the Loris Malaguzzi International Center in Reggio Emilia, covering a ground space of around 850 square meters.

The atelier activities produce various results worthy of reflection. Among other things, they make it clear how children develop theories from data collected from different sources, giving them precise meanings and reformulating their beliefs.

> M: "Perhaps the smaller tube has a picture inside it and takes it to the mirror... through a secret passage, a short-cut, and then, as quick as it can, it takes the picture to the wall."
> L: "To get to the wall, the picture in the little mirror does a run."

Thanks to the concept of image, a new mechanism, no longer inscrutable, is possible; moreover, the role of speed is significant for both children:

> L: "It's the mirror that makes it turn round the other way."
> M: "As if the comb had done a somersault."
> L: "It does a somersault on the wall."
> M: "No—first it does one on the mirror. The lens is for planning the somersault."
> L: "Mystery solved."

Marco and Lucia, through direct experience, identify the mirror-lens system, which plans and performs the rollover.

The Hundred Languages of Children

The Ray of Light atelier is inspired by the ideas of Loris Malaguzzi (1920–94), founder of Reggio Children; according to Malaguzzi children have a hundred languages and want to use them all, as explained in the following poem, "No way. The hundred is there":

*The child
is made of one hundred.
The child has
a hundred languages
a hundred hands
a hundred thoughts
a hundred ways of thinking
of playing, of speaking.*

*A hundred, always a hundred
ways of listening
of marveling, of loving
a hundred joys
for singing and understanding
a hundred worlds
to discover
a hundred worlds
to invent
a hundred worlds
to dream.*

*The child has
a hundred languages
(and a hundred hundred hundred more)
but they steal ninety-nine
The school and culture
separate the head from the body.
They tell the child:
to think without hands
to do without head
to listen and not to speak
to understand without joy
to love and to marvel
only at Easter and at Christmas*

*They tell the child
to discover the world already there
and of the hundred
they steal ninety-nine.*

*They tell the child
that work and play,
reality and fantasy,*

science and imagination,
sky and earth,
reason and dream
are things
which do not belong together.

And thus they tell the child
that the hundred is not there.
The child says:
No way. The hundred is there.

(translated by Bennett Bazalgette-Staples)

Modena, Italy   Olmes Bisi

Acknowledgments

This book would not have been possible without my participation in the Reggio Children projects. My sincere thanks therefore go to Paola Cagliari, Roberto Montanari, Giovanni Piazza, Sandra Piccinini, Carla Rinaldi, Maddalena Tedeschi, and Vea Vecchi.

I am very grateful to my colleagues and friends, Carlo Calandra, Dario Di Francesco, Lorenzo Pavesi, and Francesco Priolo for the material they provided me with and for all their valuable suggestions. Thanks also are due to Stefano Ossicini for his critical reading of the text, and to Matteo Bisi and Serena Lo Russo for their assistance on the medical topics.

Manuela Arata, Sara Di Marcello, Giovanni Filocamo, Martha Fabbri and Silvia Tagliaferri all worked professionally and effectively towards the creation of the text. This writing further benefited from the constant and valuable support of my wife.

This book, first published in Italian by Sironi Editore, was translated into English by the author, with the indispensable and qualified support of Bennett Bazalgette-Staples. Many thanks to Maury Solomon and Nora Rawn at Springer, New York, for their editorial guidance.

Contents

Preface	vii
Acknowledgments	xv
1. The History of Light	1
Pythagoras' Theorem and Sky Observations	1
The Aristarchus Method	3
Eratosthenes and the Size of Earth	6
Astronomical Computations	9
Alhazen and the Discovery of Light	11
Galileo and His Telescope Pointed Skyward	14
Kepler's Supernova	19
White Light: Pure or Impure?	21
The Eclipses of Jupiter's Moon	23
Young and the Wave Nature of Light	26
The Age of the Sun	28
The Röntgen Rays	30
The N Waves Blunder	34
The Eddington Eclipse	37
Einstein Misunderstood	41
Einstein and the Nobel Prize	43
Microwaves to the Fore	43
The Light from the Big Bang	45
2. Experiments with Light	49
Colors	49
Colored Bodies	52
Waves in Space	53
Waves in Time	56
Electric and Magnetic Fields	57
The Electromagnetic Spectrum	59
Invisible Light	61

	The Speed of Light	65
	Faster Than Light	69
	Shadows	72
	Photons	72
	The Nature of Light	75
	Irradiation and Black Body	78
	Transparency	82
	Reflection and Diffusion	84
	Mirrors	84
	Refraction	87
	Prisms	88
	Lenses	91
	Interference	95
3.	Light and the Sky	101
	The Color of the Sky	101
	Daylight	103
	Rainbows	104
	Glories	109
	Aurorae	113
	The Green Ray	116
	Lightning	120
	Earth's Atmosphere	122
	The Color of Planet Earth	124
	Space Telescopes, from Hubble Onwards	127
	The Sun	131
	Eclipses	134
	Stars and Galaxies	137
	Why Is the Sky Dark at Night?	141
	Light-Years	144
	Space Travel	147
	Antimatter	150
	Supernovae	151
	Black Holes	153
	The Active Galactic Nuclei	156
	380,000 Years Since the Big Bang	160
4.	Light and Life	165
	The Evolution of the Eye	165
	The Human Eye	169

	Color Perception	172
	Dalton and Defects in Color Perception	175
	The Grammar of Color	178
	The Color Wheel and Harmony	181
	Deceiving the Eye	185
	Mirages	188
	3D Images and the Movies	191
	Bird Sight	196
	Insect Sight	198
	Fish Sight	200
	The Sight of Some Snakes	202
	Colors and the Survival of the Species	205
	Bioluminescence	206
	Photosynthesis	210
	The Colors of Leaves	213
	The Colors of Other Worlds	214
5.	Light Techniques	217
	Solar Clocks and Sundials	217
	Overhead Projectors	221
	The Camera Obscura	224
	Image Manipulation	228
	Seeing in the Micro World	229
	Photometry	235
	The Temperature of Color	236
	Thermography	239
	Diagnostic Imaging	242
	Surgery with Images	245
	LEDs	247
	Photovoltaic Panels	251
	Optical Fibers	254
	Transmitting Through Waves	256
	Radar	260
	Satellite Positioning	262
	Fluorescence and Phosphorescence	265
	Lasers	270
	Synchrotron Radiation	273
	Photonics	279

xx   Contents

Chronology of Astronomers and Physicists
and Their Discoveries 283

Photo Credits 289

Text Quotation Credits 297

Glossary 301

Guide to Further Reading 305

Index 309

About the Author

Olmes Bisi is a professor of Physics in Engineering at the University of Modena and Reggio Emilia. He has a degree in physics with honors from the University of Modena and was a researcher at the Theoretical Chemistry Department (University of Oxford UK), at the Research Laboratory of IBM, Yorktown Heights, (NY) and at the Max Planck Institute in Stuttgart (Germany). Beside his research activities in optoelectronics, he is also involved in science education within the project Reggio Children and co-authored the book *Sociocognitive and Sociocultural Approaches to Science in Early Childhood*, Patakis Publishers, Athens, 2012.

1. The History of Light

> *"Light is everywhere, but to see it, paradoxically, we must turn it on in our minds."*
>
> Vea Vecchi, Reggio Children

Pythagoras' Theorem and Sky Observations

The first scientific laboratory was the sky, and for this reason astronomy is very ancient. Its origins and its development took place thanks to some very special conditions:

- The opportunity to look at the sky, thanks to the properties of the atmosphere, which blocks almost all electromagnetic waves but is transparent to visible light.
- The presence of regular motions of the Sun, planets and stars, phenomena that constitute a natural laboratory accessible to everyone and that, repeating over time, allows for the testing of theories.
- The human ability to create tools to make measurements, together with the capacity to discover mathematical rules for their elaboration.

The gnomon was one of the first instruments used. The motions of the shadows and of the Sun were examined by means of a simple pole fixed in the ground. The data analysis required a knowledge of the geometry of right triangles and the theorem of Pythagoras, the discovery of which dates back a long way (Fig. 1.1).

Three letters (a, b, c) that represent the lengths of the sides of a right triangle, and therefore satisfy the theorem of Pythagoras, form a 'Pythagorean triple' if the following relation holds true:

$$a^2 + b^2 = c^2$$

2   Visible and Invisible

 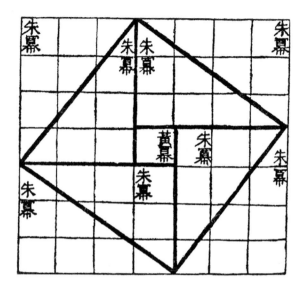

FIGURE 1.1 One of the oldest and most famous Chinese texts is the The Zhou Bi Suan Jing (Arithmetic Classic of the Zhou Gnomon). It is traditionally dated to the Zhou dynasty, i.e., the era before the first unification of the China, around 220 B.C. In this book it is demonstrated that in a right triangle with two sides of a length of 3 and 4, the third one measures 5

For example (3, 4, 5) is a Pythagorean triple, since $3^2 + 4 = 5^2$, that is, $9 + 16 = 25$.

The first Pythagorean triples, of which there are records dating back to the Babylonian civilization, are found on a partly broken clay tablet engraved with cuneiform characters containing numbers interpreted as a list of Pythagorean triples. The tablet, called Plimpton 322, has been dated to around 1800 B.C. (Fig. 1.2).

> **THE THEOREM OF PYTHAGORAS**
> In a right triangle, the sum of squares constructed on the two sides adjacent to the right angle (catheti) is equal to the square of the third side (hypotenuse) (Fig. 1.3). This theorem is a cornerstone of our entire culture, not only of mathematics. Several people have exploited it, coming up with new demonstrations. In addition to the classical discoveries of

Euclid, we may also remember those found by Leonardo Da Vinci, by the stockbroker Henry Perigal (1801–1898), by the astronomer Sir George Airy (1801–1892), and also by James Garfield (1831–1881), who later became the 20th President of the United States.

CONSIDER THIS
If there are so many possible demonstrations of the Pythagorean theorem, why are students asked to learn only one set example?

FIGURE 1.2 The Plimpton 322 tablet dates back to the reign of Hammurabi, in the first half of the eighteenth century B.C. and contains a number of Pythagorean triples

The Aristarchus Method

Estimating the distance of the Sun and the Moon from Earth is not easy, since we are dealing with distant celestial bodies. The main information that we have is their trajectory across the sky, clearly not sufficient to calculate the distance from our planet.

4 Visible and Invisible

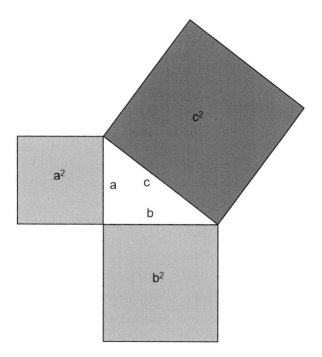

FIGURE 1.3 How many demonstrations of the Pythagorean theorem are there? Hard to say, because it is difficult to determine when a difference defines a new demonstration. However, there are many of them. The American mathematician Elisha Loomis (1852–1940) published a book in 1927 that contains 371 demonstrations of the theorem, each more or less different from the others

Aristarchus of Samos, in the third century B.C., came up with a method to assess the ratio between these distances using the information available at the time on the quadrature (see below) and eclipse of the Moon. Let's see how. He considered the positions of the Sun and Moon when the latter is in quadrature, i.e., half lit (first or last quarter). In this particular Moon phase, repeated approximately every 29.5 days, the Earth, Moon and Sun are located at the vertices of a right triangle, with the Moon on the right angle (Fig. 1.4).

Aristarchus estimated that the angle centered on the Earth was 87°, and since the sum of the interior angles of a triangle is 180°, he deduced that the one at its vertex in the Sun measured 3°.

With known angles, Aristarchus worked back to the relationship between the sides of the sky triangle thanks to a simple

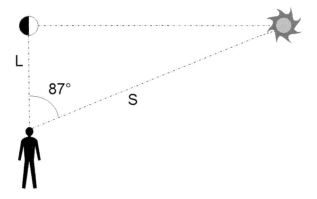

FIGURE 1.4 Estimate of the relative distances. When the Moon is in quadrature (i.e., half lighted), the angle between the Moon and Sun with the center on Earth is about 87° (in the drawing the proportions are distorted)

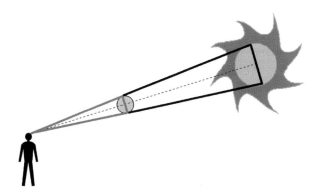

FIGURE 1.5 The calculation of the ratio between the diameter of the Sun and the Moon performed by Aristarchus is based on his observation of total solar eclipses

proportion, thus reaching the conclusion that the Earth–Sun distance is about 19 times greater than the Earth–Moon distance.

But he didn't stop there. He also observed that during a total solar eclipse, when the Sun is completely obscured by the Moon, the two discs appear to be of the same size, with the Moon totally overlapping the Sun. Considering the two triangles of Fig. 1.5, with a solid black line and a continuous gray line, he observed that the base of the larger coincides with the Sun's diameter; the diameter of the Moon represents the base of the minor, while the heights of the two triangles are nothing more than the distance of Earth–Sun and Earth–Moon.

Since he had established that the relationship between these lengths is about 19, the same was to apply for the bases. Aristarchus thus established that the diameter of the Sun is about 19 times that of the Moon.

> **ARISTARCHUS' ESTIMATE TODAY**
>
> Aristarchus' procedure was correct, but his estimate was far from the value known today, due to his excessive approximation in the measurement of the angle between the Earth–Moon and Earth–Sun directions. Today we know that the value of this angle is not 87° but 89° 51′ (where 51′ indicates 51/60 of a degree). This is a very extreme case. With only 9′ more, we would get an angle of 90°, thus an 'open' triangle, with the two longer sides parallel, and which would therefore meet...at infinity. The correct value, 89° 51′, leads us to infer that the ratio of the distances of the Earth–Sun and Earth–Moon system is not 19 but 382!
>
> So not only is the average Sun–Earth distance about 382 times that of Earth–Moon, but the same estimate applies to the relationship between the sizes of the Sun and the Moon. Today we know that the Sun's diameter is about 400 times that of the Moon.

Eratosthenes and the Size of Earth

The major center of scientific and technological studies in the ancient world, the foundation of Hellenistic civilization, was Alexandria under the Ptolemaic dynasty.

In the third century B.C., Alexandria was a rich and powerful city. On the island of Faro, facing the harbor, there stood a tower the height of which was estimated at between 115 and 135 m, with parabolic mirrors capable of sending light signals to vessels up to 50 km away. The Royal Library of Alexandria was the largest and richest library of the ancient world. Eratosthenes was its governor and made some extraordinary discoveries.

He drew the first geographical map of the then known world, using the modern concepts of latitude and longitude. His far more important result, however, was his accurate measurement of the size of Earth. It was probably an enterprise that drew on the work of various operators and state officials in a complex survey coordinated by Eratosthenes himself, as the highest scientific authority in the realm. The results of this impressive work were reported in two volumes, called *On the Measurement of the Earth*, which were eventually lost.

Fortunately, however, a simplified report survived, to be found in a text by the Greek astronomer Cleomedes, *De motu circolari coelestium corporum* (On the Circular Motions of the Celestial Bodies) dating to the first century A.D.

This popular version considers two cities: Alexandria and Syene (the modern-day Aswan) and is based on the assumption that at a place situated on the Tropic of Cancer, at noon on the first day of summer, the Sun is on the vertical and its light was even reflected in the water at the bottom of wells. It was considered that Syene was also positioned exactly on the Tropic of Cancer and that Alexandria was located on the very same meridian, although further north compared to Syene. Since at noon on June 21, the shadow in Alexandria revealed an inclination of the solar rays of 7°12′ compared to the vertical, Eratosthenes interpreted the difference as being caused by the curvature of Earth's surface (Figs. 1.6 and 1.7).

Since 7°12′ corresponds to one-fiftieth of the angle of 360° (which represents a full circle), even the Syene-Alexandria distance must be one-fiftieth of the terrestrial circumference. Multiplying by 50 the distance between the two cities, he therefore concluded that the measurement of the terrestrial circumference

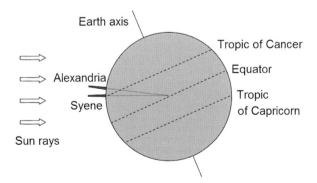

FIGURE 1.6 To calculate the length of the terrestrial circumference, Eratosthenes considered the difference between the shadows generated by sunlight at noon on the summer solstice in Alexandria and in Syene

8 Visible and Invisible

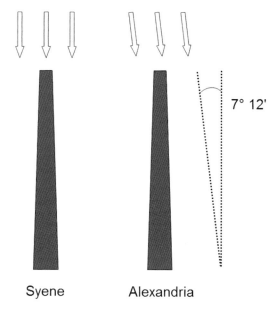

FIGURE 1.7 The solstice tilt. At noon on June 21, while in Syene, the Sun's rays strike the ground *perpendicularly*; at Alexandria they have an inclination of 7°12' from the *vertical*

was equal to 39,690 km (24,662 miles), very close to today's value of 40,030 km (24,873 miles).

After more than a century, the Greek philosopher and geographer Posidonius (135–51 B.C.) estimated the circumference of Earth to be about 28,000 km (17,398 miles), a less accurate assessment than the measurement made by Eratosthenes and its current estimate.

> **CONSIDER THIS**
> Eratosthenes and Posidonius proposed two very different estimates of the size of Earth. In your opinion, why did scholars continue to discuss the matter without actually trying to repeat the measurement until the seventeenth century? Only in 1606 did the Dutch astronomer Willebrord Snellius carry out the first modern measurements, the results of which accorded with those reached by Eratosthenes.

Christopher Columbus (1451–1506) based his project to reach the Indies on the false estimate of Posidonius, which was adopted by Ptolemy, According to his original itinerary, sailing west from Portugal would reach Asia after covering a much lesser distance than the actual one.

Astronomical Computations

In 1902, near Antikythera, a small Greek island located in the southern Peloponnese, a stone block was found with a gear mechanism inside. Further examination showed that what initially looked like a stone was actually a mechanical device consisting of 82 copper pieces dating from the second century B.C. (Fig. 1.8).

Despite the effects of corrosion, it was possible to read various inscriptions, reconstruct the mechanism's structure, and identify at least 30 gears (Fig. 1.9). The level of miniaturization and the complexity of the parts of the object, both comparable to those of mechanical clocks manufactured after the seventeenth century, aroused great surprise.

The Antikythera mechanism, which is still studied and is still source of new discoveries, turned out to be a complex mechanical device used for various functions:

- As an analog astronomical calculator used to predict the positions of celestial bodies, including the Sun, Moon and planets.
- As a Moon phase calculator, due to the differential rotation of the sidereal cycle of the Moon and the Sun's yearly cycle (anticipating by 1,500 years the discovery of differential gears, which make it possible to obtain a rotation with a speed equal to the sum or the difference of two connected rotations).
- As a calculator of calendars as proposed by Greek astronomers, based on the consideration that the lunar month lasts 29.5 days, a value not exactly contained in the year solar, which is 365.25 days long.

The pointers of the machine showed the flow of time according to different calendars, including that of Meton, a Greek astronomer of the fifth century B.C., which consisted of 6,940 days corresponding to 19 solar years or 235 lunar months. After

10 Visible and Invisible

FIGURE 1.8 The main component of the Antikythera mechanism

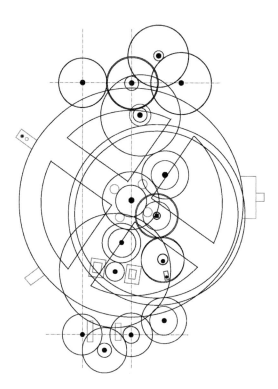

FIGURE 1.9 Through the use of X-rays, it has been possible to reconstruct the basic mechanism inside the machine

this period of time, the Moon and Sun return to virtually the same position.

The mechanism also contained an even more accurate cycle, the Callippic cycle, developed by the Greek astronomer Callippus in the fourth century B.C., covering a period of 27,759 days corresponding to 76 years, or 940 lunar months.

Finally, another pointer indicated the time by using the Saros cycle (a Chaldean word meaning "eclipses," of 223 lunar months), which contains the period of the eclipses—29 in the case of the Moon and 41 in that of the Sun. Eclipses repeat after the completion of that cycle.

In order to demonstrate that all this was possible with the technology and materials of the day, modern scientists have built a model of the Antikythera mechanism, currently residing at the National Archaeological Museum of Athens, together with the original.

Cicero (106–43 B.C.) in *De Re Publica* mentions a planetarium built by Archimedes, capable of predicting the movements of the Sun, the Moon and the five planets known at that time. It is therefore likely that the Antikythera mechanism was not unique but part of an ancient Greek tradition of complex mechanical technology, lost at the end of the Hellenistic period.

The complex apparatus of the Antikythera mechanism was reconstructed using Lego bricks and pieces, as can be seen on YouTube by typing "Lego Antikythera Mechanism." Two other videos are also available on YouTube that reconstruct the functioning and the history of the original. You find them by inputting the key words "Antikythera Mechanism Part 1" and "Antikythera Mechanism Part 2."

Alhazen and the Discovery of Light

What is light? What mechanism allows us to see objects? These are issues that have been much debated for many centuries.

In ancient Greece, there were those who, like Democritus (born between 470 and 457 B.C. and died between 360 and 350 B.C.), believed observed objects emit thin layers, called *èidola* (from the Greek *eidolon*, image), regarded as wrappers containing the shape

12 Visible and Invisible

FIGURE 1.10 Superman, the comic book hero created in 1938 by Jerry Siegel and Joe Shuster, was famous, among other things, for his so-called X-ray vision

of the object. These images entered the eye through the pupil and produced a vision (Fig. 1.10). The notion of *èidola* actually extends to the sense mechanisms of smell, based on the perception of volatile substances present in the air.

This hypothesis was challenged by various thinkers, including the Pythagoreans, Plato, and Euclid. Their theories reversed the light path, which they thought moved forth from the eye, like a lantern. The rays emitted, called "fires of vision," were described as thin wires, which departed from the eye to reach and explore objects. This concept is not unlike that of what the comic book hero Superman's X-ray vision is, since in the case of Superman, the vision also takes place with the emission of rays from the eyes.

With the demise of the Greco-Roman world, and the decline of the scientific spirit and the simultaneous diffusion of the occult sciences, much of the research into seeing was abandoned by the West. However, the Arab world preserved the knowledge from the past and continued the investigation, giving rise to highly innovative scientific research that reached its peak around the year 1000.

One of the most important scientists of that period was Ibn Al-Haytham, known in the West as Alhazen. He dealt with

FIGURE 1.11 Detail of the cover page of a Latin edition (1572) of the Book of Optics by Alhazen, which evokes the legend of the fiery destruction of the Roman fleet in front of Syracuse, featuring mirrors designed by Archimedes. In the case of the 'burning lens,' the sunlight propagates, heats, and ignites regardless of the observer

physics, philosophy, mathematics, and revolutionized optics, placing a new entity at its center—light.

Alhazen examined and rejected the theories of *èidola* and fires of vision, both born from the consideration of vision as an interaction between two entities, the object and eye. Based on several experiments, using calculation and geometry, he revolutionized the concept of vision by introducing a third element, with its own independent existence—light itself (Fig. 1.11). According to the interpretation of the scientist, the luminous beam is a mobile entity, emitted by a lit object, that propagates through different media up to the eye and which may, along its straight path, be reflected or refracted.

With this approach, the role of the eye is understood. It is an independent optical system, similar to a camera obscura, which receives the light, focuses on the image, and transmits it to the brain through the optic nerve.

With Alhazen optics and vision science was reborn. In his monumental *Book of Optics* in seven volumes, he addresses the topics of optics, physics, mathematics, anatomy, and the psychology of light. This text was to be widely used by Western scholars, in particular by Kepler, in the seventeenth century.

Galileo and His Telescope Pointed Skyward

Galileo Galilei, founder of the scientific method, was a key protagonist of the modern scientific revolution. As he himself stated in May 1609, "… a report reached my ears that a Dutchman had constructed a telescope, by the aid of which visible objects, although at a great distance from the eye of the observer, might be seen distinctly as if near." This information led him "… to inquire into the principle of the telescope, and then to consider the means by which I might compass the invention of a similar instrument." After a series of experiments, he finally built "an instrument so superior that objects seen through it appear magnified nearly a thousand times, and more than thirty times nearer than if viewed by the natural powers of sight alone."

Thus Galileo was not the first to build a telescope, although for years his was the best of those known. His revolution therefore does not lie in the invention but in the extending of the use of such a device, from mere optic wonder to an instrument of inestimable value for astronomical observation.

On pointing his telescope towards the sky, Galileo was surprised to discover wonderful new phenomena, of which he immediately understood the significance. The most important of these include the reliefs on the lunar surface, the four satellites of Jupiter, the phases of Venus, the rings of Saturn, and sunspots. His research paved the way for modern astronomy. To this day, our knowledge of the cosmos is based primarily on an analysis of the light that reaches Earth. Before Galileo, the sky was scrutinized by the naked eye alone, and the visible stars were only a few thousand; today, with a medium-sized instrument, we can observe billions of stars.

In Galileo's day, culture was dominated—especially for religious and philosophical reasons—by the geocentric system, on which astrology is still largely based. According to this view, Earth is at the center of the universe, which was its only imperfect part, due to the stain of original sin (Figs. 1.12 and 1.13).

Galileo's discoveries undermined the geocentric theory, then dominant and endorsed by the Church. For example, the discovery that four satellites rotate around Jupiter was in stark contrast with the notion that saw every celestial body orbiting Earth.

Galileo also turned his interest to the Sun, solving the problem of how to observe it without irreparable damage to eyesight, thanks to an idea of his student Benedeto Castelli (1577–1643),

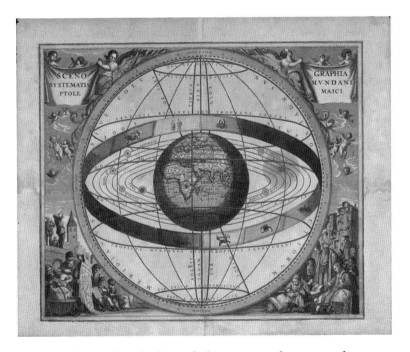

FIGURE 1.12 The perfect bodies of the Moon, planets, and stars rotate around Earth, as depicted in the Scenographia Systematis Mundani Ptolemaici by Johannes van Loon (1611–1686) and shown here. Thus, the Moon was represented as "translucent, solid, firm, and polish'd bright, like adamant," as Dante states in Canto II of Paradise, despite the presence of dark patches visible to the naked eye. Significant in this respect is the exchange between Dante and Beatrice in Paradise regarding the Moon (same canto), in which the poet asks his guide about the much-discussed origin of lunar spots: "But tell, I pray thee, whence the gloomy spots upon this body [...]?"

16 Visible and Invisible

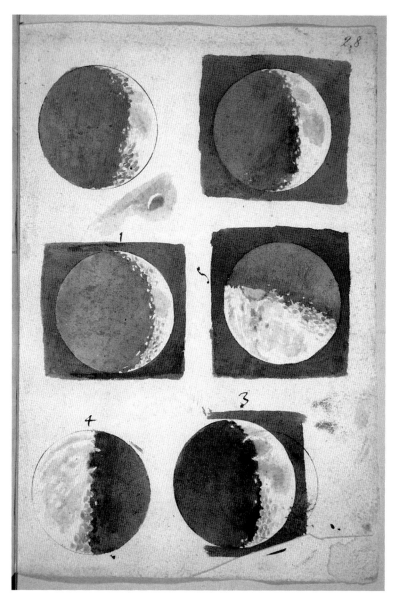

FIGURE 1.13 The telescope showed to Galileo the details of the lunar soil and the origin of the spots due to a deeply rippled surface, which produced, among other observable shadows, reliefs similar to those of the imperfect Earth. Galileo reported these observations in various drawings, like the one shown here, which dates back to November/December 1609 and represents the phases of the Moon (Galileo's original drawing is in the National Central Library of Florence)

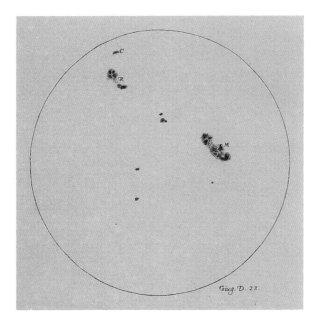

FIGURE 1.14 For the Jesuit and German mathematician Christoph Scheiner (1573–1650), a perfect body like the Sun had no spots, and those visible even to the naked eye were no more than shadows, arising from planets moving between Earth and the Sun. Galileo participated in the dispute about the origin of these spots, by showing 35 drawn tables, one per day, documenting the evolution of a group of spots. Since these observations were made at the same time of day, the drawings directly showed the motions of the spots and their rise and dissolution on the solar disk. Shown here is the original drawing of sunspots observed and documented by Galileo on June 23, 1613

who suggested projecting the image onto a blank sheet of paper, so as to be able to study it for a long time and at any time of day, without danger. The most important study performed in this way was that of sunspots, which Galileo was able to accurately observe even when the Sun was very high in the sky (Fig. 1.14). Thus he disproved the hypothesis that these spots were due to shadows cast by other heavenly bodies onto the solar surface and placed in doubt the description of the Sun as a perfect and unchangeable star. All this by pointing a telescope toward the sky, without preconceptions and with an open mind.

> **PLANET EARTH: STATIONARY OR IN MOTION?**
> Galileo discovered what is now universally known as the Galilean principle of relativity, that is, the physical equivalence of inertial reference systems. We apply this law to the case of a ship moving at a constant speed in a perfectly calm sea. No passenger, in an environment without an external view, can tell whether the boat is stationary or moving. As Galileo believed that the speed of circular motion of celestial bodies was constant, as in the case of the ship, it was impossible to verify whether Earth was stationary or rotating around the Sun.

The similarities of the rotations around Earth and those around the Sun disappear if we consider the motion of other celestial bodies. The orbits of the planets are very simple when referred to the Sun, while they become very complex if they are considered relative to Earth. In the Ptolemaic system, which was geocentric and based on the perfect circle, the motion of the planets took place based on the first circumference, the center of which rotated along a second one, thus describing the complex geometric form of the epicycle. This curve, however, is only an approximation of the actual motion of the planets as seen from Earth. In fact, the orbits of celestial bodies are not circular but elliptical, and only the heliocentric Copernican model was able to incorporate this modification easily.

The divergence between Galileo and the Catholic Church, which led to the deplorable treatment of the scientist, was about whether Earth and humanity were at the center of the universe or not, with the incorruptible heavenly bodies revolving around it. Galileo challenged a closed system of knowledge, based on books and authorities of the past, and supported the validity of knowledge obtained through the direct investigation of nature. Einstein wrote: "The leitmotif which I recognize in Galileo's work is the passionate fight against any kind of dogma based on authority. Only experience and careful reflection are accepted by him as criteria of truth." For this reason, Galileo is considered the founder of the modern scientific method.

Kepler's Supernova

On the night of October 9, 1604, a new bright object appeared in the sky in the constellation of Serpentarius (Fig. 1.15). Its light was similar to that of a star and was visible to the naked eye for a period of 18 months, then disappeared again into nothing.

What was the origin of this new star? According to the dominant (Ptolemaic) cosmological model, cosmic objects were perfect and immutable. How could a star be born and then die? Galileo pointed out that this event was therefore in conflict with the hypothesis of the celestial spheres as unalterable systems.

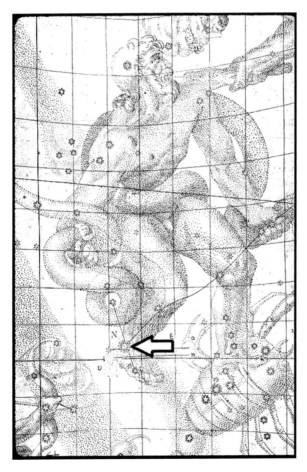

FIGURE 1.15 An original drawing by Kepler depicting the constellation of Ophiuchus with the SN 1604 marked with the letter N (indicated by the *arrow*)

20 Visible and Invisible

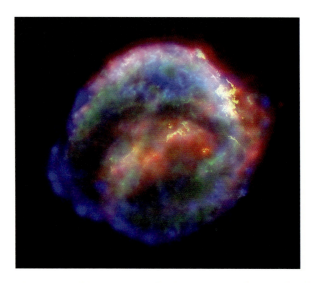

FIGURE 1.16 To create this image, which gives an idea of the distribution of Kepler's supernova remnant, three space telescopes were used. The X-rays were collected by the Chandra X-ray Observatory, the visible radiation was recorded by the Advanced Camera for Surveys (ACS) mounted on the Hubble telescope, while the infrared band was collected by Spitzer

This phenomenon turned out to be very special. It was not the birth of a star but its death. The observed object, located in our galaxy about 23,000 light years away, was a supernova, the final stage of the evolution of a massive star leading to a catastrophic explosion, with an enormous emission of light, a billion times that of the Sun. The supernova was cataloged with the abbreviation SN 1604 and is thought to have been the last to be observed in the Milky Way.

The German astronomer Johannes Kepler studied this star for a long time and published the results of his observations in the book *De stella nova in pede serpentari* (On the New Star in Ophiuchus' Foot). For this reason, the supernova was named Kepler's Star.

The remains of Kepler's Star are visible to this day, more than 400 years since the appearance of the supernova. It's an enormous sphere of dust and gas that expands at a speed of 6 million km (about 3.6 million miles) per hour and that at present occupies a region some 14 light-years wide.

Figure 1.16 shows an analysis of the supernova remnant obtained by collecting not only visible light but also X-rays and infrared radiation. The maps of the different radiation types are converted into visible images, assigning different colors to various signals—blue for high-energy X-rays, green for low-energy X-rays, yellow for visible light, and red for infrared.

> **CONSIDER THIS**
> The explosion of the star and the formation of the supernova did not happen in 1604. In that year the light given off from this catastrophic event reached Earth. Since the distance that separates us from the supernova remnant has been estimated at 23,000 light-years, the light took about 23,000 years to reach us, and so the explosion occurred when we were still in the Stone Age. Can that be true?

White Light: Pure or Impure?

White light, that of the Sun *par excellence*, was considered the very essence of light for centuries—it's parts inseparable but simple and homogeneous, its metaphorical connotations of purity intrinsically connecting it to the image of God.

By contrast, the various colors were thought to be created by colored bodies, and were considered distinct from white light. The formation of the rainbow, due to the prism, was described as the inclusion of color in light, consequently modified.

In 1672, Isaac Newton informed members of the Royal Society that he had carried out an experiment that demonstrated beyond doubt that sunlight, known as white light, was not pure, as had been believed until then, but was made up of a mixture of colors. Newton called his study the *Experimentum Crucis* (Crucial Experiment)—a milestone in the history of science as well as proof of the validity of the experimental method.

In the *Experimentum Crucis* the prism does not generate new colors but simply those already contained within white light, which is not altered but rather broken down. Hence, the traditional view is reversed. The colors are simple and homogeneous, while white light is a mixture of various components (Figs. 1.17 and 1.18).

22 Visible and Invisible

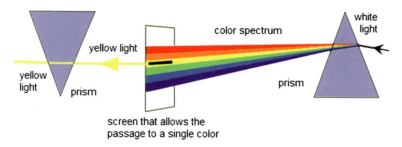

FIGURE 1.17 The Experimentum Crucis. Newton shone a ray of *white light* through a prism to reveal the colors; then he made the spectrum hit a screen where there was a small opening that let through a ray of monochromatic light, *yellow* for example, which then passed through a second prism. If it was the prism that colored the light, on exiting the second prism this beam should have changed color. But it was not so. Newton thus showed that the colors were the components of *white light*, and that the prism was limited to merely making them visible

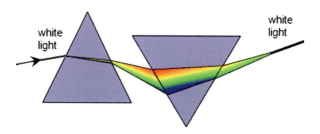

FIGURE 1.18 In a second experiment, Newton confirmed his hypothesis about *white light* by recreating it as a mixture of different colors, placing two prisms with one side parallel to the other, the first to create the rainbow spectrum, the second to create different colors. To better understand the construction of colors, it may be useful to think about the extreme case in which the distance between the two disappears and their parallel faces join. The two merge to become a single body with parallel faces, and thus a ray of light passes through without generating a rainbow

CONSIDER THIS

In the publication sent to the Royal Society in 1672, Newton tells of his surprise at the form of the beam. Rays sent through a circular hole, on passing through the prism, produce a very elongated shape on the wall. Why?

Light, according to Newton, consisted of 'corpuscles' that travel into space at great speed. A specific mass may even be attributed to each color, the greater towards the red end of the spectrum and the smaller towards the violet end. Newton foresaw seven different masses, therefore seven primary colors, the others being obtained from their combination. When light encounters a body, be it via reflection or refraction, an interaction takes place through the universal law of gravity discovered by Newton himself, that bodies interact with light through the force of gravity acting on the particles, and their different masses explain the different refractions of the colors.

> **CONSIDER THIS**
> What is the main difference between Newton's corpuscles and Einstein's photons? Unlike photons, corpuscles are particles with their own mass. Therefore, if we cross two light beams, assuming that light is composed of particles, we expect a broadening of the rays, due to the collisions between the particles. In the case of photons, on the other hand, the beams intersect without any broadening. Einstein's model is actually more in line with reality because the light rays, while crossing, remain unchanged. Can that be true?

The Eclipses of Jupiter's Moon

Soon after the discovery of the four main satellites of Jupiter, Galileo proposed to use their motion to accurately measure the passage of time. Such a clock would have been able to solve a fundamental problem in timekeeping: the determination of the longitude of a place. Galileo's idea was not successful and was put aside until 1650, when it was applied by cartographers and topographers for drawing more accurate maps than before. Consequently, the need arose to have accurate tables outlining

FIGURE 1.19 This picture, sent June 25, 1979, from the Voyager 2 spacecraft, shows Jupiter's moon Io against the giant planet around which it orbits. In this photo, details of up to only 200 km (about 120 miles) across can be made out. Galileo called the four satellites of Jupiter that he has observed the 'Medicean stars' in honor of Cosimo II de' Medici, Grand Duke of Tuscany. Along with Io there were Ganymede, Callisto and Europa

the motions of Jupiter's satellites. To this purpose, around 1670 a group of astronomers from the Paris Observatory, led by the Italian Giovanni Cassini (1625–1712) and including the young Dane Ole Rømer, carried out very precise observations on the motion of the largest of Jupiter's satellites, Io (Fig. 1.19).

Io performs a complete orbit of Jupiter in about 42.5 h. The easiest way to measure the duration of the orbit is to record its eclipse by reporting the moment in which the satellite enters Jupiter's shadow cone or emerges from it.

The motion of Io around Jupiter is periodical, like the Moon around Earth. Surprisingly, however, it was realized that it was not regular. In certain periods of the year the satellite emerged from the dark side of Jupiter later. Giovanni Cassini, perhaps on behalf of the entire group of astronomers, was the first to propose an explanation for these discrepancies: light travels at a finite speed and it takes different times to reach here, depending on the distance between Jupiter and Earth.

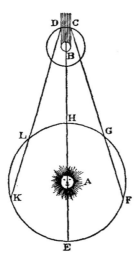

FIGURE 1.20 In the original design that appeared in the Rømer's publication, the Sun is denoted by A and Jupiter by B. The shadow of sunlight cast by the planet is also shown. When Io, which rotates around Jupiter, is in its shadow cone, in positions ranging between C and D, it is not illuminated by sunlight and thus eclipses. Earth moves around the Sun, and during the year reaches different positions (E, F, G, H, L, K). We examine the measurements of Io's motion on the basis of its disappearances in C. On days when Earth is at point F, the cycle is about 3 min longer than that measured when our planet reaches point G. The light travels along different paths and therefore the time required to cover the distance varies

Cassini later changed his mind and rejected this solution, while giving free reign to the ever more convinced Rømer to pursue the hypothesis. The 32-year-old Danish astronomer published his results in the *Journal des Sçavans* (Journal of Savants), one of the first modern scientific journals (Fig. 1.20). Indeed, he is rightly regarded as the one who provided proof of the finite speed of light. On the basis of these considerations, Rømer estimated that time taken by light to travel along the diameter of Earth's orbit was about 22 min; the modern estimate is 16 min and 40 s.

The first to use these data to evaluate the speed of light was the Dutch physicist Christiaan Huygens, who obtained an estimate of 230,000 km/s (≈143,000 miles/s) against the 300,000 km/s (≈186,000 miles/s) considered correct today.

Young and the Wave Nature of Light

Nature is perplexing, and sometimes individuals are born to pursue its study with relentless determination. Thomas Young, born in 1773 into an English family of Quakers, was definitely one of them. He learned to read at age 2, and at 4 years of age had already read the Bible twice. Later he came to know as many as 14 languages, becoming a respected physician and a great scientist (optics, mechanics of solids, physiology) not to mention a prominent Egyptologist, being among the first to discover the method for deciphering hieroglyphs.

At that time, the nature of light had not yet been unambiguously determined. According to Newton and his followers, light was made up of very small particles, emitted in all directions from a luminous body. Other scientists did not share the Newtonian corpuscular model, being convinced that the light was somehow made up of waves, similar to those of a liquid or a sound (Fig. 1.21).

This latter hypothesis was due to some peculiar properties of light. For example, the Italian scientist and Jesuit Francesco Maria Grimaldi had discovered that a light beam, after passing through a small hole, did not keep its shape but broadened, illuminating also the part that was supposed to be in the shade, insofar as it was unreachable by direct rays. The same extension of the radius also occurred for waves of liquid and sound. Grimaldi first described this with the term 'diffraction.' Even Huygens had proposed elements in favor of the wave model of light, but still lacked definite proof and so was unable to convince everyone.

This proof was ultimately provided by Young, through the experiment illustrated here and known as Young's Interference Experiment. A light beam was shone in the direction of a wall with two slits (see Fig. 1.22).

If the light had been composed of corpuscles, one should have been able to only see the two strips indicated in (b). Instead, the result of the experiment was surprising, for as is shown in (c), several strips appeared, some very bright, separated by non-illuminated strips, giving a set of light and dark bands that formed an 'interference pattern.'

FIGURE 1.21 In a liquid, the phenomenon of wave interference is clearly visible, as can be seen by throwing a few stones into an expanse of still water and observing the ripples they produce on the surface

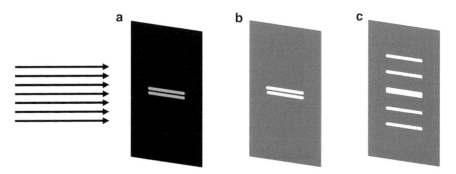

FIGURE 1.22 Young's interference experiment. The interference pattern that formed in (**c**), beyond the two slits in the screen (**a**), provided definite proof for the wave theory of light. The expected result according to the corpuscular model is shown in (**b**)

The light therefore showed its wave nature, the same easily observable in a liquid. Waves have a maxima and a minima, and when two of them meet, the result is maximal in some regions and minimal in others. In the case of light this means an alternation between bright and dark zones.

The Age of the Sun

Where does the light radiated from the Sun, which makes life possible on Earth, come from? How long has this energy source been operating? How old is the Sun?

These are closely related questions that can be difficult to answer. Suppose that our star is powered by a chemical reaction, such as that which occurs in a gas-burning boiler. By taking a quantity of reagents equal to the mass of the Sun, it could generate heat for a good 3,000 years. That may seem like a long time, but it's not enough to clear our doubts, for the age of this celestial body is certainly far greater.

During the nineteenth century, many scientists asked themselves the same questions. At that time, the only alternative mechanism able to explain the energy source of the Sun was based on Newton's law of universal gravitation. Bodies are attracted to one other, or in other words, they gain energy on approaching each other. But where does this end? We also find this in thermal energy—a compressed body heats up.

An evaluation of the age of the Sun was provided by one of the greatest physicists of the time, one of the founders of modern thermodynamics, the Englishman William Thomson (1824–1907), who was later to become Lord Kelvin based on scientific merit (Fig. 1.23). According to the scientist's reasoning, the gravitational energy of the Sun, accumulated during its formation, could account for the production of light for a time equal to 20 to 40 million years, giving a new estimate of the Sun's age. As a consequence, the age of life on Earth could not be greater than 20 to 40 million years. In the same period, Charles Darwin (1809–1882) published the first edition of *On the Origin of Species by Means of Natural Selection*, suggesting the evolution of plant and animal species by natural selection—a slow process that requires hundreds of millions of years (Fig. 1.24). Based on geological studies, Darwin estimated the age of Earth to be around 300 million years.

The conflict between the findings of the two great scientists was intense. If the age of the Sun was that estimated by Kelvin, the time available was not sufficient for the development of evolution. Kelvin had no doubts about his evaluation, but Darwin

FIGURE 1.23 A portrait of the English physicist Lord Kelvin, who estimated the age of the Sun according to its gravitational energy

was so impressed by the arguments of the physicist that in later editions of *On the Origin of Species*, he eliminated any reference to the timescale of evolution.

Today we know that Darwin was right and Kelvin wrong. It is estimated that the Sun and Earth are of a similar age, around 4,600 million years, a value compatible with the evolutionary process.

Kelvin's error was due to the limits of physics at that time. Today the theory of relativity provides the correct approach to understanding the origin of the Sun's energy. Einstein's famous relationship between mass and energy, $E = mc^2$, in which $c^2 = 9 \times 10^{16}$ m^2/s^2, shows that a small amount of mass can be converted into an enormous amount of energy. This is what happens in the Sun, where the processing of hydrogen into helium takes place. In the reaction, part of the nuclear mass disappears and energy appears in its place.

30 Visible and Invisible

FIGURE 1.24 Charles Darwin (here photographed by J. Cameron in 1869) was one of the protagonists in the debate on the age of the Sun and Earth

> **THE TIMES PERIODS OF EARTH**
> The timescale of our planet is divided into intervals of different sizes: eons (billions of years), eras (hundreds of millions of years), periods and epochs.

The Röntgen Rays

In November 1895, a 50-year-old professor of physics at the University of Würzburg and newly appointed rector, Wilhelm Röntgen continued his investigation into the mysterious cathode rays, then an important research topic. A high electrical voltage in a vacuum tube generated effects that were invisible, yet detectable by the fluorescent material deposited inside a glass tube (Figs. 1.25 and 1.26).

The History of Light 31

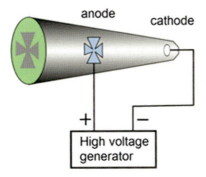

FIGURE 1.25 Diagram of a vacuum tube, known as a Crookes tube, used by Röntgen when he discovered X-rays. The cathode, the small flat disk to the *right*, was connected to the negative pole of the generator, while the anode, at the center and shaped like a Maltese cross, was connected to the positive pole

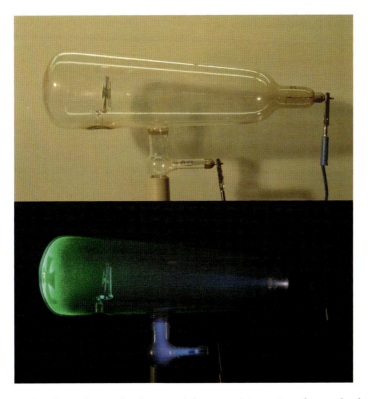

FIGURE 1.26 When the cathode rays (electrons) in a Crookes tube hit the *right* end of the tube, it emitted *green light*, by fluorescence. The shadow created by the anode demonstrated the *straight line* motion of the cathode rays

During his experiments, Röntgen darkened the laboratory room and placed the apparatus in a thick-walled black box, preventing the light produced inside to escape; then to his great surprise, he saw fluorescent wording light up, about 2 m from the box.

It was known that cathode rays were unable to travel such a long way. Why, then, was the writing illuminated? No known phenomenon could explain it!

Röntgen shut himself in his laboratory for 6 weeks and eventually solved the riddle. The cathode rays hitting the glass generated a new type of radiation that can pass through solid objects, like the box isolating the tube. The new rays, mysterious at the time and so called X-rays, were able to pass through many substances. Since the soft parts of the human body were transparent to these rays, while metals and bones appeared opaque, these rays, being able to leave a trace on photographic plates, made it possible to 'photograph' the denser parts inside the human body (Figs. 1.27). 'Radiography,' exploiting X-rays, spread rapidly, also due to the mistaken belief that such radiation was harmless.

Today we know that X-rays are nothing more than high-frequency electromagnetic radiation. Their energy makes them potentially dangerous for living beings, as they are able to 'ionize' atoms and thus cause damage to chains of DNA or cells. When used with caution, however, they are useful and sometimes essential in certain fields of study, such as medicine.

CATHODE RAYS
Röntgen discovered X-rays while investigating the nature of the cathode rays, and because of this discovery in 1901 won the first ever Nobel Prize for Physics.

The nature of cathode rays was clarified in 1897 by the English physicist Joseph Thomson (1856–1940), who identified them as electron beams, which he called 'corpuscles.'

The History of Light 33

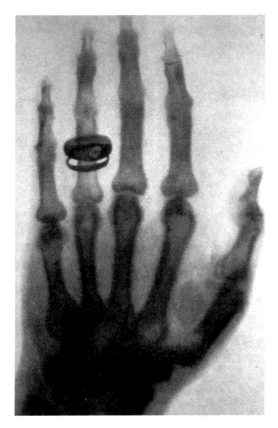

FIGURE 1.27 The first public presentation of the 'unknown rays' took place on the evening of January 23, 1863, in front of an influential audience gathered at the regular meeting of the Society of Medical Physics of Würzburg. At the end of his speech, Röngten asked the famous professor of anatomy Albert von Kölliker permission to perform a picture of his hand using X-rays. The radiograph that materialized in front of the eyes of those present, shown in this image, soon became famous. Kölliker proposed that the new rays be called Röntgen rays

X-RAYS AND SYNCHROTRONS
The wavelength of X-rays is very short, with a length comparable to that of atoms (nanometers). X-rays for this reason are also used to investigate atomic structures that nanoscience and nanotechnology have developed. The production of these X-rays, however, no longer takes place using glass tubes, but through large and complex machines called synchrotrons.

The N Waves Blunder

Two different kinds of invisible rays, cathode rays and X-rays, were discovered within a short time. Shortly after the identification of X-rays by Röntgen, the French scientist René Blondlot (1849–1930) announced a third type of invisible radiation: N-rays.

Blondlot had previously studied radio waves and was the first to accurately measure their speed, which turned out to be that of light.

According to Blondlot, N-rays were able to penetrate aluminum, to make a filament coated with fluorescent calcium sulfide shine, and to increase the brightness of an electrical spark. In three years, nearly 300 studies by over 100 scientists were published, mostly French, on this discovery. Its author received awards and honors. According to the press of the time, humanity was riding a new wave, perhaps more important than that of X-rays (Fig. 1.28).

Many scientists, however, found themselves unable to reproduce Blondlot's results. Among these was the American Robert Wood (1868–1955), a pioneer in the study of ultraviolet radiation, another type of invisible wave. Wood decided to go to France and visit Blondlot's laboratory to analyze his experimental procedures closely. Such visits, to this day, are widespread in the scientific community and are born out of the need to check whether a result is reproducible. In the French laboratory, therefore, the evidence and the characteristics of N-rays were shown to the visiting scientist.

During the first demonstration, the N-rays were directed towards an electrical spark, the brightness of which should have increased. Wood saw no effect and was told that his eyes were not sensitive enough.

Subsequently, several photographs were presented showing the brightness of the spark, with and without N-rays, but Wood verified that they were unreliable, as they were produced in conditions given to cause various errors (Fig. 1.29). Finally, the decisive proof was presented to him. The rays were generated in a tube, then deflected by an aluminum prism towards a thin fluorescent wire that glowed slightly when it was hit by them. All this was done in a dark room to visualize the lighting effect, which was very weak. The absence of light allowed the American scientist,

FIGURE 1.28 The cover of a book by René Blondlot on N rays

36 Visible and Invisible

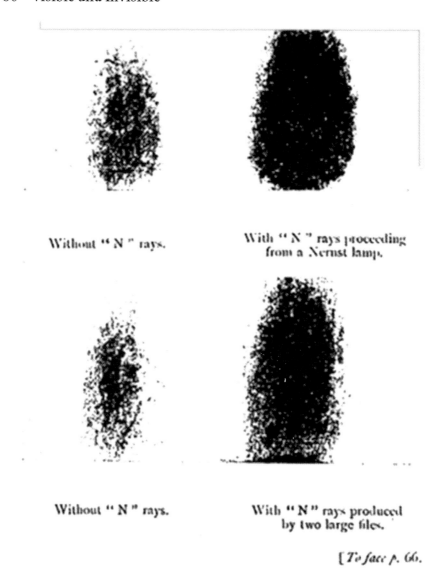

FIGURE 1.29 The images shown to Robert Wood in France, demonstrating the variation in the brightness of a spark in the presence or absence of N rays

whose suspicions were growing, to remove the prism, unbeknownst to Blondlot and to his assistant. Without this component, the machine could not function. In spite of that, the French physicists claimed to see the effect of N-rays. Soon after Blondlot repositioned the prism, but his collaborator saw him bustling and thought that he was removing it. In the experiment that followed, the assistant could not see any ray, even though he should have, for the equipment was supposedly fully operational!

Wood published the results of his investigations in the journal *Nature*, suggesting that N-rays were a purely subjective phenomenon and that scientists had been presented with data corresponding to their expectations, without any real confirmation. N-rays then disappeared from the scientific literature, although Blondlot proceeded in his university activities, continuing to believe in their existence.

The Eddington Eclipse

The Theory of General Relativity proposed by Einstein in 1916 changed the concept of gravity. It was no longer due to a force acting at a distance (Fig. 1.30) but to the fact that mass *bends* the space surrounding it (Fig. 1.31).

The new theory looked more powerful, but it had to be verified experimentally. The traditional Newtonian approach accurately described the motions of bodies and planets; a new proposal might replace it only if it were found to be able to properly address unexplained phenomena with the previous system. Since the two theories described the motions of everyday objects with such small differences that they could not be verified, it was necessary to consider extended bodies such as stars and planets in order to detect any discrepancies.

In 1916, Einstein stated that general relativity explained the anomalies of the orbit of Mercury, the closest planet to the Sun and the one that rotates around it at the highest speed. Other explanations had been proposed, but the relativistic one did not have to resort to any special assumptions.

However, it was not the crucial proof, capable of convincing scientists of the validity of the new theory. Einstein had long since identified another way to do it. According to his calculations, light is deflected by the 'curvature' of space. Light rays coming from a group of stars, which reach us after having brushed past the Sun,

38 Visible and Invisible

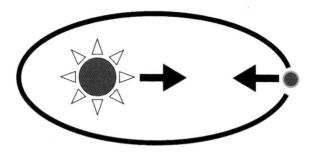

FIGURE 1.30 According to Newton, Earth rotated around the Sun due to the gravitational force acting between the Sun's mass and that of Earth

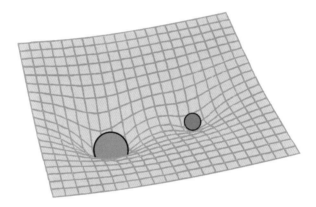

FIGURE 1.31 According to Einstein, the curvature of space around the Sun causes Earth's rotation

change direction due to the curvature. Therefore their image is different from that which we receive at night, when the rays do not pass close to the Sun (Fig. 1.32).

The expected deviation is very small; indeed, the light beam rotates about 0.0005°, difficult but not impossible to measure.

Observing stars so close to the Sun in full daylight appeared practically impossible, except during a total solar eclipse, in which the Moon completely obscures the sunlight, making even those stars whose rays pass alongside the Sun visible. Therefore, to observe such a phenomenon, several expeditions were organized.

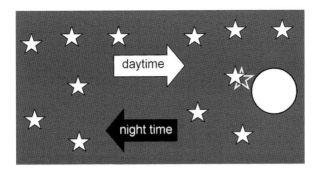

FIGURE 1.32 According to Einstein's theory, the image of the same group of stars appears to us differently when viewed at night or during the day, when their light passes very close to the Sun. Such diversity is due to the deviation caused by light rays emitted by the celestial bodies because of the curvature of space

The first mission went to Argentina in 1912, but it was not possible to view the eclipse due to cloudy skies and rain. Another attempt was made in 1914 with a German expedition that set out for the Crimea. This time the failure was caused by the outbreak of the war, which prevented observation. Again, due to the war, it was not possible to organize an expedition to Venezuela on the occasion of the 1916 eclipse, and 2 years later, an American attempt to measure the effect did not provide conclusive results.

In 1919 the British organized two expeditions: one to Sobral in Brazil, the other to the island of Principe in Africa. In southern South America the weather was nice, but the results were ruined by a flaw in the telescope, while the expedition to Africa, led by astronomer Sir Arthur Eddington, found the sky covered with clouds and was able to take only a few decent photographs. After long and careful analysis, however, these images showed that Einstein was right: the light had deviated by the value predicted by the theory of relativity (Fig. 1.33).

Scientists and the entire world now realized that a profound revolution in science had taken place, with a new theory of the universe. On November 10, 1919, *The New York Times* ran the headline: "Lights All Askew in the Heavens. Men of Science More or Less Agog Over Results of Eclipse Observations. Einstein Theory Triumphs."

40 Visible and Invisible

FIGURE 1.33 One of the photos taken during the 1919 African expedition led by Eddington. The stars positioned around the Sun during the eclipse belong to the famous and bright cluster of the Hyades, the head of the animal being represented by the constellation of Taurus. The astronomer highlighted them with a dash, which was then compared with that obtained from a night image of the Hyades

GENERAL RELATIVITY TODAY
Very accurate measurements were performed using radio waves from distant sources, the quasars. These confirmed Eddington's results without any doubt. Today GPS systems, which assist us in journeys, also use the theory of relativity to correctly identify our position.

> **CONSIDER THIS**
> During the long journey then necessary to cross the Atlantic by ship, Eddington calculated the total number of charged elementary particles with great precision (protons or electrons) in the universe. This number, eighty digits long, is called the 'Eddington number' and is considered the largest number in nature. Can that be true?

Einstein Misunderstood

The possibility that light is composed of photons was formulated by Albert Einstein. In 1905, at the age of 26, the great physicist published fundamental works in three different fields:

- The nature of light with the hypothesis of the photon, or "quantum of light," in the study of the photoelectric effect.
- The real existence of atoms through the explanation of Brownian motion.
- The electrodynamics of moving bodies, with the Theory of Special Relativity.

The idea of the photon was certainly the discovery by Einstein that was most resisted by the scientific community. In 1913, some of the greatest German scientists, including Max Planck and Walther Nernst (1864–1941), proposed that Einstein should be appointed as a member of the Prussian Academy. Their presentation expressed the highest appreciation for his scientific work. Significantly, however, it was concluded as follows: "In sum, one can say that there is hardly one among the great problems, in which modern physics is so rich, to which Einstein has not made a remarkable contribution. That he may sometimes have missed the target in his speculations, as for example in his hypothesis of light-quanta, cannot really be held too much against him for it is not possible to introduce really new ideas even in the most exact sciences without sometimes taking a risk (Fig. 1.34)."

The American physicist Robert Millikan (1868–1953), known for having determined the electric charge of the electron, and for

42   Visible and Invisible

FIGURE 1.34 Albert Einstein at the age of 42, photographed by Ferdinand Schmutzer. The image was taken at an event in which the scientist spoke to an audience of 3,000 people. A witness described the event as follows: "The public was in a strangely excited state, in which it no longer mattered whether one understood what was being presented, but that one was in close proximity to a place where miracles occur"

his measures on the photoelectric effect, wrote in 1916: "Despite.... the apparently complete success of the Einstein equation [for the photoelectric effect] the physical theory of which it was designed to be the symbolic expression of [was] found so untenable that Einstein himself, I believe, no longer holds to it."

Thirteen years after the proposal of the photon, in 1918, the discoverer of the theory of relativity wrote in a letter to his friend Michele Besso: "I do not doubt anymore the reality of radiation quanta, although I still stand quite alone in this conviction."

Einstein was right and all the others were wrong! Today we can state and verify with extreme accuracy that the light really is composed of photons. Not only that, the studies of Einstein showed new properties of electromagnetic waves. Among these we might remember stimulated emission, in which light is emitted under the influence of the incident radiation, a phenomenon that was to contribute to the birth of the laser.

Einstein and the Nobel Prize

Seventeen years after his study of the photoelectric effect and the proposal of the photon, in 1922, Einstein was awarded the 1921 Nobel Prize for Physics with the following explanation: "for his services to theoretical physics, and especially for his discovery of the law of the photoelectric effect."

Note that it speaks broadly of the contributions to theoretical physics without mentioning the theory of relativity or the discovery of the photon. It just indicates the law of the photoelectric effect, the validity of which everyone agreed on. Regarding the explanation, Abraham Pais (1918–2000), scientist and biographer of Einstein, writes: "This is not only an historic understatement but also an accurate reflection on the consensus in the physics community."

Microwaves to the Fore

Percy Spencer (1894–1970) was a researcher at the American company Raytheon, which, among other things, specialized in the manufacturing of a fundamental constituent of radar, the valve generating the microwaves. This device, named the magnetron, was widely used during World War II, giving the Allies a significant advantage over the Nazi and Japanese armies and thus influencing the course of the war.

In 1941, Spencer made an important improvement to the device, modifying it so as to render it simpler. Thanks to this innovation it was possible to increase the daily production of valves from 17 to 2,600.

One day, towards the end of the war, he was working around an operating magnetron when he felt something strange. A chocolate bar in his pocket had melted!

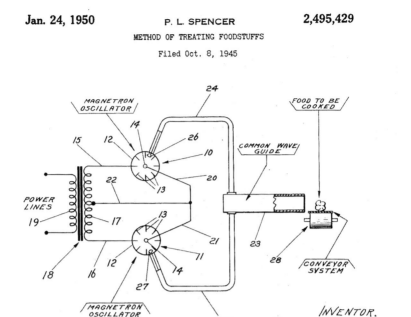

FIGURE 1.35 Patent No. 2,495,429, for the invention of the microwave oven

His curiosity aroused, he put some popcorn in front of the magnetron and it began to crackle all over the room. Finally, he put a raw egg in a pot in front of the magnetron. The egg exploded, splattering a colleague, giving final confirmation that the 'microwaves' generated in the device could cook food quickly and unconventionally.

In 1946 Raytheon patented the microwave cooking process, and the following year the first commercial microwave oven, called Radarange, was produced. It was nearly 6 feet tall and weighed in at 750 pounds (Fig. 1.35).

Today this type of cooking is widespread. The microwaves used have a wavelength of 12.24 cm and carry little energy, less than that of the infrared rays that transmit most of the heat in traditional ovens (Fig. 1.36). Neither microwave nor infrared waves are able to ionize the substances cooked and therefore cannot damage the genetic material (DNA) in cells so they do not make food any more likely to cause cancer.

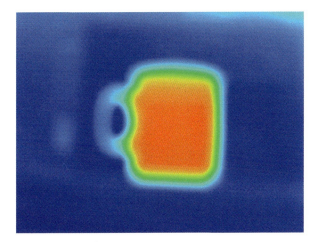

FIGURE 1.36 Infrared photograph of a cup full of water heated by a microwave oven. It is clearly seen that the cup remains relatively cold compared to the liquid inside

> **CONSIDER THIS**
> While being cooked in a microwave oven, food is rotated. Why?

The Light from the Big Bang

Earth's atmosphere absorbs most of the electromagnetic waves from the celestial bodies, except those corresponding to visible waves, microwaves, and radio waves. For this reason the investigation of the cosmos also makes use of antennas that collect signals from the microwave range.

In 1961, upon completing his doctorate in microwave physics, a young Arno Penzias was recruited as a researcher at the Bell Telephone Laboratories in New Jersey. His project was to continue his studies in radio astronomy. This was research of a theoretical nature, with no direct application, as was frequently the case in major industrial laboratories at the time.

In 1964, thanks to a series of fortunate circumstances, one of the antennas used by the Bell Telephone Company for satellite

46 Visible and Invisible

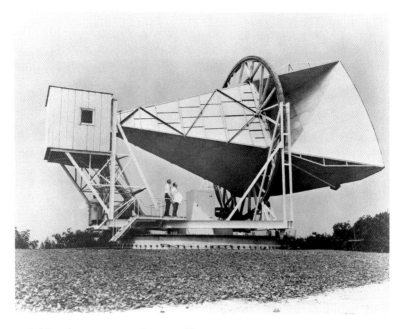

FIGURE 1.37 The antenna for satellite communications of the Bell Telephone Company, adopted by Penzias and Wilson

communications became available for Penzias' research, who had meanwhile been joined by the radio astronomer Robert Wilson. The instrument was a huge horn-shaped antenna, capable of detecting microwaves better than any other system in existence (Fig. 1.37). Before starting measurements it was necessary to fine tune the device, eliminating all the disturbances to better detect weak signals from space. There were many possible sources of interference, from radio and radar waves to the electromagnetic noise generated by the mechanisms inside the antenna itself. Penzias and Wilson were able to tackle and solve all these problems. However, there remained a faint background noise that they could not remove. Every possible source of interference was examined, but with no result. They finally decided to make a careful inspection of the device itself. They found a pigeon's nest and droppings inside the antenna. Could these droppings be the reason for the background noise? The device was cleaned and, as might be expected, the noise decreased. However it did not disappear.

Not far from them, at Princeton University, a group of scientists were designing an instrument able to detect cosmic background radiation that, according to cosmology, must permeate the universe as a result of the Big Bang. After examining the characteristics of the noise detected by Penzias and Wilson, they realized they were sitting on the astronomical discovery of the century. The signal received corresponded to radiation coming uniformly from all directions of space, emitted by a black body with a temperature of $-270\ °C$, just $2.7\ °C$ above absolute zero, the lowest possible temperature.

According to the Big Bang theory, the initial singularity produced a homogeneous level of radiation, evidence of the event itself, which after a cooling period of billions of years, corresponds to a temperature a few degrees above absolute zero.

The signal detected by Penzias and Wilson was in perfect accordance with these theoretical predictions. For the first time, someone had been able to experimentally test hypotheses about events that happened 14 billion years ago. As Ivan Kaminov said, one of Penzias' and Wilson's colleagues, "They looked for dung but found gold, which is just opposite of the experience of most of us."

The studies on cosmic background radiation led to the assignment of two Nobel Prizes for Physics. The first was awarded to Penzias and Wilson in 1978 for their discovery; the second went to George Smoot and John Mather, for their studies on anisotropy, the weak dependence of the cosmic microwave background radiation on the direction of observation, an analysis carried out through the use of the COBE (Cosmic Background Explorer) satellite.

2. Experiments with Light

"I have to understand the world, you see."
Richard Feynman, scientist and Nobel laureate

Colors

Colors are a feature of the light that we humans perceive. They identify the wavelengths of the visible radiation contained in luminous rays. Some animals are sensitive to different electromagnetic waves than we are. For example, a bee's eye can detect ultraviolet light but not red light.

The various hues can be mixed in two ways: either by adding different lights or subtracting certain colors from a light beam.

Let us begin with additive synthesis. By mixing two colored lights, we get a third one. With only three color elements—red, green, and blue—we can obtain the full range of colors (RGB model). These three hues are called primary because by combining equal amounts of them you get white, the sum of all colors.

White is not really a color but a mixture of colors. White light can be broken down using various techniques, for example by using a prism.

What happens if we add equal amounts of two primary colors? We get the secondary colors: yellow (red + green), cyan (green + blue), and magenta (blue + red) (Fig. 2.1).

Yellow and blue, which add up to make white, form a pair of 'complementary' colors. The other complementary pairs are cyan–red and magenta–green. The eye uses additive synthesis to process light signals, and the functioning of devices such as video cameras and displays relies on the same technique.

However, there is another way to mix hues—subtracting color from color (subtractive synthesis). This happens for example when we look at an object through sunglasses with lenses that

50 Visible and Invisible

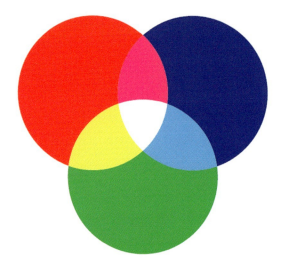

FIGURE 2.1 Partially overlapping one *blue* luminous ray, one *green*, and one *red* one, and projecting them onto a neutral background, you get the secondary colors of additive synthesis (*yellow, cyan, magenta*) as well as *white*

absorb only certain wavelengths. When different-colored filters are superimposed, each subtracts part of the light, and the final color is the result of all the removals.

Subtraction also occurs by mixing together different-colored substances. To obtain the desired color, a painter applies several layers of paint, which are like a series of filters. In this way, new colors are formed.

Subtractive synthesis may be considered complementary to the additive one. The primary colors of subtraction are the secondary ones of the additive synthesis—cyan, magenta, and yellow (CMY model). Also, in this case, secondary colors are obtained by mixing two primaries. The blending of magenta and yellow is equivalent to a filter that lets only red pass through it. Likewise, cyan + yellow gives green, and cyan + magenta makes blue.

By mixing equal amounts of yellow, cyan, and magenta you get black, which is the absence of color (Fig. 2.2).

The composition of colors is not a large field of science, but it is crucial in many activities, primarily in painting. In this context, an approach has been developed akin to that of additive and

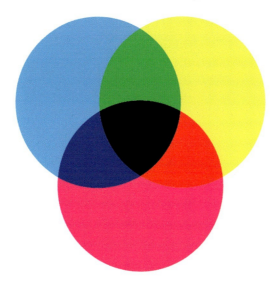

FIGURE 2.2 Observing a light beam through a *yellow* filter, a *cyan*, and a *magenta* one, partially overlapping, you get the secondary colors of subtractive synthesis (*red, blue, green*) as well as *black*

subtractive synthesis as defined in physics, but with one important difference: the three primary hues are red, yellow, and blue. This triple has proved useful both in additive and subtractive synthesis, although red, yellow and blue do not produce white in the former and black in the latter.

COLORS AND WAVES

What are colors? Nothing but the manifestation of an electromagnetic wave, and in particular of its wavelength, λ, which, in the visible region of the spectrum is expressed in microns (μm), i.e., or millionths of a meter.

The shortest wavelength of the visible spectrum (0.4 μm) corresponds to violet; the longest one (0.7 μm) to red; and intermediate values to the remaining colors. Below violet ($\lambda < 0.4$ μm) we have ultraviolet radiation, while those above red ($\lambda > 0.7$ μm) are classified as infrared rays.

TABLE 2.1 The visible spectrum

Wavelength (µm)	Color
0.400–0.450	Violet
0.450–0.500	Blue
0.500–0.570	Green
0.570–0.590	Yellow
0.590–0.620	Orange
0.620–0.700	Red

Colors are not objective properties of light but are subject to the processing and interpretation carried out by our eye-brain system. For example, brown does not match any wavelength. It is a barely luminous yellow-orange, simply interpreted by the brain as a different color. The visible spectrum is continuous and presents no jumps when passing from one color to another. Nevertheless, it is possible to establish ranges for the various colors to a good degree of approximation (Table 2.1).

Colored Bodies

The color of an object depends on its response to illumination. In general, when we perceive a certain color it means that the portion of the light corresponding to it is reflected or scattered and the remaining portions are absorbed (Fig. 2.3).

Black objects completely absorb luminous rays striking their surface, while white ones reflect or diffuse them entirely. If a body is transparent, its color corresponds to the light that is able to cross it without being absorbed.

In metals, light remains confined to the surface layers, and thus another effect may occur. Most of the waves striking it are reflected from the surface, and little light penetrates beyond. If the metal surface is flat, the light beam is reflected in a single outgoing direction, and so the metal behaves as a reflective surface. Mirrors in fact consist of a thin layer of metal placed beneath a layer of glass.

Experiments with Light 53

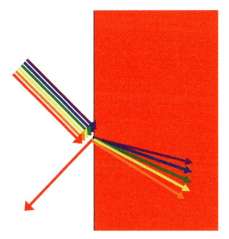

FIGURE 2.3 White light striking a *red* body. *Red* light dominates the spectrum diffused by an opaque *red* body, while the other colors are largely absorbed by the material. The way in which an object appears to us not only depends on its intrinsic color but also on other factors, including the context, the color of the light, its intensity, and the environment between the observer and the object. For example, the color of the Sun varies according to the layers of the atmosphere crossed by the light emitted

> **CONSIDER THIS**
> In winter, dark clothing, which absorbs the heat from sunlight better, is worn more often, while instead it is reflected from white clothes, typical of the summer months. The same principle applies to animal fur. If this is true, why does the polar bear have white fur?

Waves in Space

Several wave phenomena are familiar to us: the oscillation of sea water, the vibration of air exposed to sound, and so on. Light also behaves like a wave. When we consider light, the oscillating movement is that of the electric and magnetic fields, shown respectively in black and gray in Fig. 2.4. For this reason, light is considered an electromagnetic wave.

54 Visible and Invisible

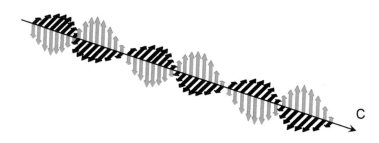

FIGURE 2.4 The electric (*black*) and magnetic (*gray*) fields and the propagation speed (*black arrow*) of an electromagnetic wave

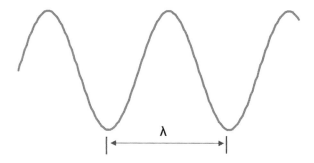

FIGURE 2.5 The electromagnetic wave varies in space and time. To examine its spatial dynamics, imagine photographing the electric (or magnetic) field of a wave. In the snapshot that we obtain, every variation refers only to space, time being fixed. Spatial distribution is described by wavelength λ, which specifies the distance between two peaks or troughs of the oscillation

Electromagnetic waves do not need any support medium to propagate; therefore, light can even travel in a vacuum, and on Earth we can receive waves emitted by the Sun, other stars, and even galaxies. Its speed of propagation, which identifies the direction of motion, is perpendicular to both the electric field and the magnetic one, and in the top figure is indicated by the black arrow and the symbol 'c' in Fig. 2.4.

How can waves be measured? From the point of view of their oscillation in space, we talk about their wavelength (Fig. 2.5), which is measured using the meter, its multiples and submultiples. The most widely used multiple is the kilometer (Table 2.2):

$$1 \text{ km} = 1,000 \text{ m} = 10^3 \text{ m}$$

Experiments with Light

TABLE 2.2 The most important submultiples of the meter

Name	Symbol	Value	In powers of 10
Millimeter	mm	1 thousandth of a meter	10^{-3} m
Micrometer	μm	1 millionth of a meter	10^{-6} m
Nanometer	nm	1 billionth of a meter	10^{-9} m

The wave just shown is 'polarized,' because the direction in which each field oscillates is always the same. Normally, light sources emit streams of radiation, each independent from the next. The resulting radiation is not polarized because the fields oscillate in ever-changing directions.

> **CONSIDER THIS**
> Often the reflection of the sky on the water surface hinders clear vision (Fig. 2.6a). However, if the water surface is observed with a pair of glasses equipped with polarized lenses, the reflection is eliminated and the image is clearer (Fig. 2.6b). The reflection on the water surface polarizes the light, which is then stopped by the filter of the glasses. Can that be true?

FIGURE 2.6 (a) The sky's reflection on the water surface. (b) The same water surface photographed using a polarized lens. The main difference lies in the reflected light of the *central right* area and *upper right* corner

Waves in Time

A light wave varies not only in space but also in time. Imagine filming its electric (or magnetic) field at a certain point in space. Its variation may be described by the period T, the quantity measured in seconds, indicating the duration of a complete oscillation (Fig. 2.7).

The change in time of a wave is similar to that in space. In the first case the range of variation is determined by the period T, in the second case by the wavelength λ.

To describe the temporal variation of a wave, instead of the period T the frequency (ν) is often used, equal to the inverse of T:

$$\nu = 1/T$$

Frequency indicates the number of complete oscillations that a wave performs in a second and is measured in cycles per second or hertz (Hz), named after the German physicist Heinrich Hertz (1857–1894), who carried out major research into the field of electromagnetism. The main multiples of hertz are indicated in the following table (Table 2.3).

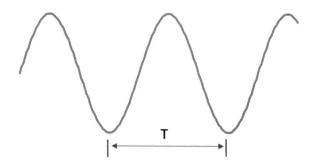

FIGURE 2.7 To examine the temporal dynamics of a wave its period must be assessed, i.e., the time interval necessary for the wave to perform a complete oscillation

TABLE 2.3 The multiples of Hertz

Name	Symbol	Value	In powers of 10
Kilohertz	kHz	1,000 Hz	10^3 Hz
Megahertz	MHz	1 million hertz	10^6 Hz
Gigahertz	GHz	1 billion hertz	10^9 Hz
Terahertz	THz	1,000 billion hertz	10^{12} Hz

Electric and Magnetic Fields

In our daily activities we often deploy forces, for example, when we move objects or raise weights. Normally these actions require direct contact between the one deploying the force and the body to which it is applied. Moving objects at a distance, perhaps with the power of thought, is an old dream that has never become a reality.

Nature, however, is full of surprises and is able to deploy forces without the need for direct contact, through 'action at a distance.' Let us consider a magnet, for example. Taken individually it behaves like any other object, but if two magnets are brought together, these will repel or be attracted to each other even without being placed in contact, since a force is created even at a distance (Fig. 2.8).

Action at a distance raises several questions. What is the invisible link that connects the two magnets? If you move one of them, does the second one sense the change immediately or after a given length of time?

To answer these and other questions, the notion of the magnetic field was introduced. A magnet, even alone, creates a field. The field is invisible, but it alters the properties of the space around it and contains energy—in this case, magnetic energy.

The action at a distance between two magnets can thus be described as mediated interaction. Each magnet is subject to a force because it lies in the field generated by the other (Fig. 2.9).

When we consider a charged particle instead of a magnet, we get a similar situation, yet the concept of the magnetic field is replaced by that of the electric field. A field of this kind

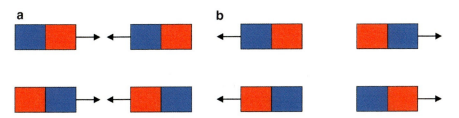

FIGURE 2.8 (a) If we bring together the opposite poles of two magnets the magnets attract each other. (b) If we bring together equal poles they repel each other

58 Visible and Invisible

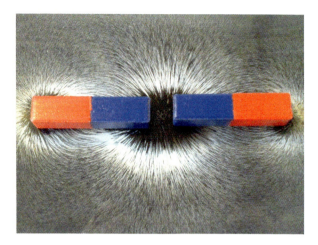

FIGURE 2.9 The magnetic field around two magnets repelling each other, detected using iron filings dipped in oil

cannot have an autonomous existence. A constant field cannot exist in isolation but follows the magnet or electric charge that generated it.

If, however, we can create a field that varies in time, for example by oscillating electric charges, it generates a new phenomenon. The motion of charges creates electric and magnetic waves that move away from the source, self-supporting themselves. The energy carried by this radiation is distributed in equal parts in the electric or magnetic form.

The fields created in this way propagate at high speed even in empty space, through electromagnetic waves. The waves lead a life independent from the source that generated them, and so electric or magnetic fields can be investigated just like any material object. Albert Einstein wrote in this regard: "The electromagnetic field is, for the modern physicist, as real as the chair on which he sits."

The Scottish scientist James Maxwell was able to bring together all previous observations, experiments, and equations related to electricity and magnetism in a single theory. His work led to the formulation of the four Maxwell equations, which describe the behavior of the relative fields and are capable of predicting all electric or magnetic phenomena.

Furthermore, Maxwell realized the possibility of creating a new type of wave, the electromagnetic wave, using oscillating electric and magnetic fields. He was even able to calculate the speed at which these waves propagate and, using only electrical and magnetic properties (at that time known only approximately), obtained a value close to that of the speed of light. In this regard, he wrote: "This velocity is so nearly that of light, that it seems we have strong reason to conclude that light itself [...] is an electromagnetic disturbance in the form of waves propagated through the electromagnetic field according to electromagnetic laws." How right he was!

The Electromagnetic Spectrum

Visible light does not cover the entire range of electromagnetic waves. There are other waves of the same nature that are invisible to us. The range of all these waves, including light, constitutes the electromagnetic spectrum.

What distinguishes the various portions of the spectrum? The only difference is in the ν frequency. As it varies, it produces different types of radiation, from radio waves (with ν being only a few to a few hundred Hz) to gamma rays. Visible light occupies a small portion of the spectrum, bordering infrared and ultraviolet waves (Fig. 2.10).

Although frequency is an intrinsic characteristic of electromagnetic waves, at a constant frequency, the wavelength varies according to the medium that the wave passes through. Wavelength and frequency vary in opposite ways; when the former increases, the latter decreases. The long wavelength of radio waves corresponds to low frequencies, while gamma rays are characterized by very short wavelengths and high frequencies.

The properties of each wave type vary on the basis of how they are produced, and how they may be detected therefore also varies. Waves may even be produced by different means but in the same part of the spectrum. For example, waves with a wavelength of 12.2 cm produced by the magnetron of a microwave oven form microwaves, while those of 12.2 cm emitted by a Bluetooth antenna may be classified as radio waves.

60 Visible and Invisible

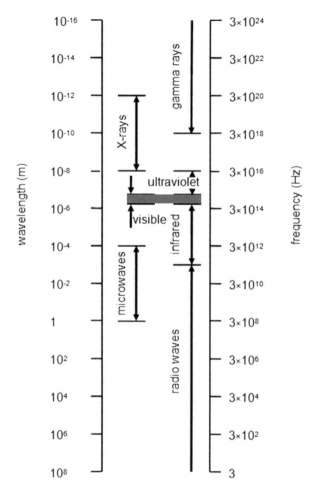

FIGURE 2.10 The various parts of the electromagnetic spectrum

There is a fundamental relationship between wavelength and frequency. Their product provides the wave propagation speed. As a consequence, by multiplying the wavelength (λ) and frequency (ν) of an electromagnetic wave, we obtain the speed of light, which in a vacuum is indicated with the letter c:

$$\lambda \nu = c$$

For example, red light has a wavelength of:

$$\lambda = 0.7 \mu m$$

and a frequency of:

$$\nu = 4.3 \times 10^{14} \, \text{Hz}.$$

Therefore the final output is:

$$\lambda\nu = 0.7 \, \mu\text{m} \cdot \left(4.3 \times 10^{14}\right) \text{Hz} =$$
$$= \left(0.7 \times 10^{-6}\right) \text{m} \cdot \left(4.3 \times 10^{14}\right) \text{cycles/s} =$$
$$= (0.7 \cdot 4.3) \times 10^{8} \, \text{m/s} =$$
$$= 300{,}000{,}000 \, \text{m/s} = 300{,}000 \, \text{km/s}.$$

We have thus calculated the value of the speed of light and indeed of all electromagnetic waves in a vacuum. This value is very high, but when electromagnetic waves pass through matter, the speed of propagation is reduced. In this case, the wavelength decreases proportionately, while the frequency remains unchanged.

We speak of 'infrared' and 'ultraviolet' radiation by referring to their frequency, $\nu = 1/T$. In this case, the terms infrared (from the Latin *infra*, meaning 'below') and ultraviolet (from the Latin *ultra*, 'beyond') indicate radiation with frequencies respectively less than that of the color red and greater than that of the color violet.

Invisible Light

When the frequency of electromagnetic radiation is outside the range of the visible, the human eye does not detect any luminous effect; however, this does not mean the absence of radiation but rather the presence of invisible light. An example is that of infrared rays, not visible because they are at wavelengths greater than those of the color red (Fig. 2.11).

Ultraviolet (UV) rays are instead characterized by wavelengths shorter than those of violet (0.4 µm) and longer than those of X-rays, with the extreme ultraviolet positioned around 0.01 µm, equivalent to 10 nm. When estimating the effect of UV rays on human health, three ranges are considered, according to different wavelengths: UVA (from 0.4 to 0.315 µm), UVB (from 0.315 to 0.280 µm), and UVC (from 0.280 to 0.100 µm).

The light emitted by the Sun contains ultraviolet radiation that, fortunately for us, reaches Earth's surface only in small

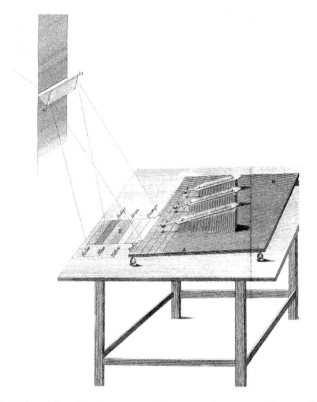

FIGURE 2.11 Infrared radiation was discovered in 1800 by William Herschel as a result of a study on the properties of sunlight, broken down into the different colors by a prism. Herschel intended to study the heating properties of the different colors by using thermometers and placing them in the area illuminated by the prism. To his great surprise, he observed that heating also took place in the areas that were not illuminated, beyond the red section, thus revealing the existence of invisible—infrared—rays, capable of heating the thermometers. The image shows the scheme of Herschel's experiment, as published in the Philosophical Transactions of the Royal Society of London in 1800

amounts. Earth's atmosphere in fact absorbs about 98% of it, and the remaining 2% that reaches us mainly consists (99%) of UVA rays.

If absorbed in moderate amounts, UVA rays are useful for the human body. They help in the prevention of various diseases, including rickets and multiple sclerosis. Ultraviolet light also allows us to identify bacteria or fungi that would otherwise be invisible (Figs. 2.12 and 2.13). Many birds are also able to see ultraviolet light, as are some insects such as bees.

FIGURE 2.12 Wood's lamp (or 'black light') is a device that emits UVA rays and, to a lesser extent, visible light. It is used to illuminate fluorescent and phosphorescent materials, for example, in the fight against the counterfeiting of banknotes and documents, which often incorporate symbols or designs only visible under ultraviolet rays

Ultraviolet radiation with frequencies higher than those of UVA, like the UVB and UVC waves, are dangerous; however, UVC waves, being a germicide, can be used in the disinfection of environments and contaminated objects.

At frequencies higher than those of ultraviolet radiation, we have X-rays and gamma rays, whose photons are even more energetic than those of ultraviolet waves.

The energy transmitted by some ultraviolet light (UVB, UVC, and extreme ultraviolet), by X-rays and by gamma rays is so high as to become ionizing, i.e., capable of extracting electrons from atoms, changing their chemical bonds and thus altering biological molecules.

Ionizing radiation is dangerous to living species even when irradiation, that is, the amount of radiation emitted, is low. In fact, it is possible for the DNA of damaged cells not to be repaired properly. This may lead to cell mutation and therefore to tumor formation. However, this is a rare occurrence, depending on the amount of ionizing radiation received and resulting in effects of the type 'all or nothing.'

64 Visible and Invisible

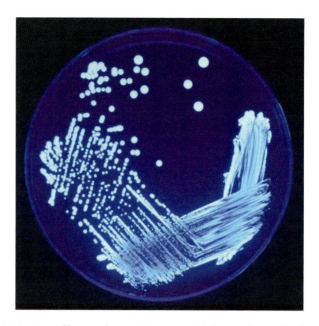

FIGURE 2.13 Legionella are bacteria not visible to the naked eye but are detectable using ultraviolet rays

X- and gamma rays from the cosmos are largely absorbed by Earth's atmosphere. In spite of this, ionizing radiation is present in our environment, mainly because of radioactive substances in Earth's crust. There is therefore natural 'background' ionizing radiation that can be detected anywhere on Earth, and it varies from place to place depending on the composition of the rocks found in a particular location.

Since it makes use of ionizing radiation, a radiograph constitutes a risk to the patient—limited but not absent. For this reason it should only be performed if strictly necessary, on the basis of an evaluation of the pros and cons. The usefulness of the examination, for the purposes of health care, must outweigh the risk (Fig. 2.14).

Can we estimate the risk of a typical radiograph? Yes, through comparison with that of irradiation due to natural background radiation, considering its average value. The irradiation to which we are subjected in chest radiography is equivalent to what we receive from natural background radiation in a month. By using the units that measure the radiation dose, this quantity is equal to 0.2 millisieverts.

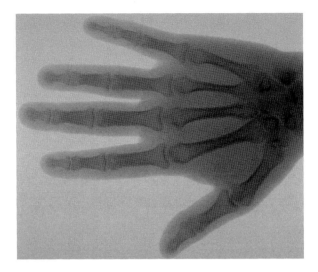

FIGURE 2.14 X-rays are used extensively in medical imaging, which has the task of identifying pathologies by examining images of the internal organs of the human body

The Speed of Light

Light has some unique features that make it special. These properties are attributes of all electromagnetic waves, of which light is a part. First of all, its incredible speed, so high that it was once thought infinite, is based on the fact that the transmission of light signals seemed instantaneous. Today we know that light travels at 300,000 km/s, i.e., about 1 billion km per hour. It is hard for us to comprehend such a speed, denoted by the symbol 'c,' from the Latin word *celeritas*, which means 'speed.'

This is the speed at which light travels through the cosmos. It takes just over one second to travel from the Moon to Earth, and eight minutes to reach us from the Sun.

The universe is so big that even a ray of light takes a long time to cross it. The Hubble Space Telescope has obtained images of a galaxy so distant that the light took 13 billion years to reach it! This image, captured in 1996, left the remote galaxy when the universe had been in existence for just 1 billion years. From the speed of light, a measure of cosmic distances can be derived, the light year, i.e., the distance light travels in one year. The galaxy, photographed by the Hubble telescope, is 13 billion light years away.

66 Visible and Invisible

FIGURE 2.15 A Ferrari 599 GTO reaches a maximum speed of 335 km/h. Light is more than three million times faster

FIGURE 2.16 The speed of sound in air is 1,235 km/h. Light is almost 900,000 times faster

Another particular property of light is that it has zero mass; indeed, it consists of massless particles called photons that, for this very reason, can travel at such speed. Any object with mass, however, cannot reach that speed (Figs. 2.15, 2.16, 2.17, 2.18, and 2.19).

Experiments with Light 67

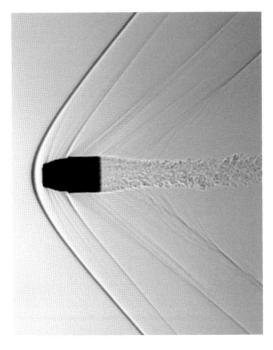

FIGURE 2.17 A bullet can be fired at a speed of 1,908 km/h. Light is more than 500,000 times faster

FIGURE 2.18 The maximum speed reached by a Lockheed Blackbird is 3,529 km/h. Light is more than 300,000 times faster

68 Visible and Invisible

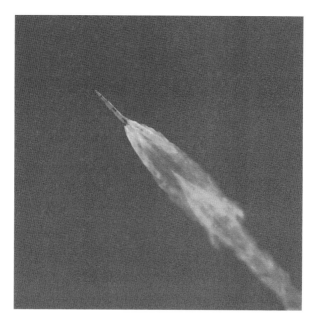

FIGURE 2.19 On May 16, 1969, Apollo 10, during its return voyage, reached a speed of 39,897 km/h. Light is 27,000 times faster

GRAVITATIONAL LENSING

Light has no mass and no weight yet is affected by the force of gravity. According to the theory of relativity, when light passes near a large mass (a star or a galaxy), its path is deflected by gravity. This change of direction has the same consequences as the deviation caused by a lens. It can focus or distort the image. For this reason we speak of 'gravitational lensing.' The curvature depends on the size of the mass. It may be that this is so high as to fold back the beam, preventing it from continuing and deflecting the radiation toward the star. The case described is that of the black hole, a star with a mass so large that it can absorb everything, even light.

> **NEW HYPOTHESES ON THE SPEED OF LIGHT**
> Einstein's theory of relativity assumes that the speed of light in a vacuum is a universal constant of nature. This is the basis on which our description of physical phenomena is founded. Recently some scientists, including Giovanni Amelino-Camelia at La Sapienza, University of Rome, and João Magueijo at Imperial College, London, have proposed alternative hypotheses, still a matter of lively debate among experts. According to these new theories, called VSL (Varying Speed of Light), velocity 'c' has not remained constant throughout the evolution of the universe. Several unanswered cosmological questions could be given new solutions on the basis of the following assumption—that in the wake of the Big Bang, light traveled at speeds much greater than it does at present.

Faster Than Light

Is it possible to travel faster than light?

This question is not a mere curiosity. The speed of the electromagnetic signals (visible light, radio waves, etc.) is the factor that limits the effective exploration of the cosmos. In the universe there are approximately 100 billion galaxies, entities of enormous size ranging from dwarf galaxies (with a few tens of millions of stars) to giant ones (including as many as trillions of stars).

The great Andromeda Galaxy, near to us and visible even to the naked eye, lies about 2.5 million light-years away. The light we see coming from it does not show us the present image of the galaxy, however, but the way it was 2.5 million years ago, when the rays of light departed. In the case of the most distant galaxies, we may see the light emitted by them up to 13 billion years ago!

Nature is full of surprises. For all we know, nothing prevents the existence of objects that can travel faster than light. The laws of physics, particularly the theory of relativity, allow us to

70 Visible and Invisible

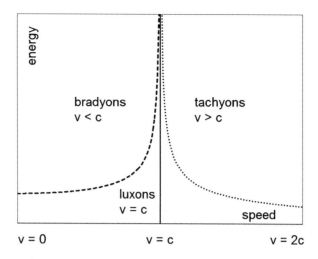

FIGURE 2.20 The graph shows the relationships linking the speed of bradyons (*dashed line*), luxons (*solid line*) and tachyons (*dotted line*) with the energy needed to reach such speeds. If we infer that the speed of luxons is always c, we may deduce that an infinite amount of energy is required to accelerate bradyons up to the speed of light and to slow tachyons down to the same speed

distinguish three different types of behavior with regard to the speed of light, c:

- that of bodies that may have different speeds but always less than c. This concerns normal matter, of which we are made and which forms the objects that we know. Normal matter has a certain inertia, and in order to increase its speed it needs energy (dashed curve in Fig. 2.20). Objects of this type are called bradyons, i.e., slow-moving entities, from the Greek word *bradýs*, which means 'slow.' When a bradyon approaches the speed c, the energy needed for such motion becomes infinite; hence it is never able to achieve that speed;
- objects that travel at the speed of light. These are particles that do not possess mass and the solid curve in Fig. 2.20 shows that their velocity is c whatever their energy. These bodies are called luxons, from the Latin word *lux*, light. They do not accelerate or stop, and the only way to slow them down is to destroy them. Photons belong to this category of particles;

- entities that may have different speeds, but always higher than that of light. There is no law of nature against this hypothesis. Starting from the assumption that anything that is not prohibited may exist, we can admit the possible existence of particles called tachyons, i.e., objects that move quickly, from the Greek word *takhýs*, which means 'fast.' The characteristic of the tachyon is indicated by the dotted curve in Fig. 2.20. There is no upper limit to the speed of tachyons; indeed, at infinite speed, the energy required for their motion becomes zero, and it increases when we try to restrain them. To slow a tachyon down to the speed of light, however, infinite energy is required. It is for this reason that these particles can only go faster than c.

Because we cannot break through the barrier of the speed of light, c is a kind of watershed between the slow bradyons and the fast tachyons.

The hypothesis of having objects traveling at speeds above c is quite attractive, but raises considerable problems, such as that of being able to influence the past. Let's take an example. Consider that Mr. Smith and Mr. Jones are located in two different places, are in relative motion, and can transmit 'tachyon signals,' i.e., traveling at speeds greater than light. Smith sends one of these messages to Jones who, after having received it, responds with a second message, sent even faster. Well, the answer might reach Smith even before that he has sent the first letter!

The possibility of influencing the past traveling at speeds exceeding that of light inspired the following limerick, published in the British satirical magazine *Punch* in 1923 by the biologist Reginald Buller (1874–1944):

There was a young lady named Bright,
Whose speed was far faster than light;
She started one day
In a relative way,
And returned on the previous night.

Actually, it is not possible to influence past events. A careful examination of the properties of tachyons has shown that these conditions cannot exist, and thus they cannot be used to send information or particles at speeds greater than that of light. Tachyons may therefore exist, even though they have not yet been detected, but they would not be able to help us with superluminal transmissions. We must therefore limit ourselves to seeing them in action in the movies or in science fiction novels.

Shadows

How fast can a shadow move? If you cast the shadow of your finger onto a wall and then shake it, the projection moves faster than the finger. As we move away from the wall, the speed of the shadow increases with the increasing distance, even though the movement is delayed because the light takes longer to cover the distance (Fig. 2.21). Over great distances from the source, our shadow will go so fast as to exceed c. This does not conflict with the theory of relativity, since the shadow does not carry energy. Literally nothing is transferred. In other words, no object can travel faster than light, but the shadow is not an object. It is the absence of light. We could assert that, even though they do not have energy, shadows can transmit information, and in this way these travel at superluminal speed. But this assertion would be wrong. Shadows are not able to provide communication faster than c because their position can only be detected when the light arrives. It is therefore possible for something to move faster than light, but not to make things travel at speeds greater than c particles or information.

Photons

So far we have described light as a wave. In actual fact, it is a very complex phenomenon. Although it often behaves like a wave, sometimes it appears completely different, as if the beam was formed by a stream of particles—photons (Fig. 2.22).

The photon is a rather peculiar particle belonging to the family of luxons. It always moves at the speed of light ($c = 300,000$ km/s), without acceleration or slowing from the moment it is emitted. Nothing with mass can travel faster. All other bodies, from the car to the electron, behave differently. Their speed may vary, even reaching high values, but always less than c. This uniqueness of light is due to its mass, the quantity of which dictates its resistance to acceleration. The mass of the photon is indeed zero.

Experiments with Light 73

FIGURE 2.21 Shadow of the plume due to the launch of the space shuttle projected by the Sun (behind the photographer) onto the Moon. The launch took place at sunset on February 7, 2011

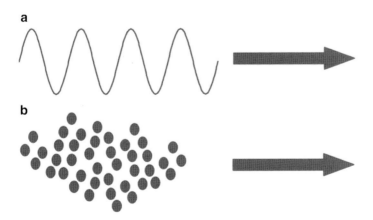

FIGURE 2.22 The dual nature of light. Sometimes a ray of light appears to us as a wave (*above*), and in other cases as a flow of photons (*below*). In the figures, where the *arrows* indicate the direction of the light beam, it is clear that the two descriptions are completely different and hardly attributable to the same entity

All this is true in a vacuum, where luminous rays, such as those that come from the stars, always travel at 300,000 km/s. When light passes through matter, its speed slows to values lesser than c. For example, the value with which it propagates through glass is equal to 200,000 km/s.

> **PHOTON ENERGY**
> Each photon is associated with a frequency (ν) and an energy (E). These two quantities are not independent but closely related. Specifying the energy of a photon is equivalent to fixing its frequency; conversely, indicating the frequency of a luminous beam implies determining the energy of its photons.

The relationship between energy and frequency is very simple. They are directly proportional, i.e., their ratio is constant. To determine the value of energy, it is sufficient to multiply the frequency by the fixed value of this ratio.

The constant defined thus is called Planck's constant; it is denoted by the letter 'h' and is a fundamental constant of nature: $E = h\nu$.

The higher the frequency of a photon, the greater the energy transported by it. In the microscopic world, the unit of measurement of energy is the electron volt (eV). The red photon (i.e., that of electromagnetic radiation corresponding to red light) has the energy of approximately 1.6 eV, while that of violet exceeds 3 eV; infrared has lower energy, down to $1/000^{th}$ of an eV. As the frequencies decrease, so do energy levels. The photons of a radio wave at a frequency of 100,000 Hz (i.e., with a wavelength $\lambda = 3$ km) have an energy of 0.4 billionths of an eV.

As we move towards the higher frequencies, we find the X photon, the energy of which varies from 100 to 100,000 eV, while the greatest energy is that of gamma rays, which may be greater than a million eV. Under particular conditions, a gamma photon with an energy level of at least 1,022,000 eV meets a strange fate. It annihilates itself (disappears) and in its place two particles appear—an electron and a positron (antielectron)—both endowed with mass.

How is it possible that an object without mass may produce a pair of particles that have mass? This can happen because the energy (E) of the photon is converted into mass (m), according to Einstein's famous formula $E = mc^2$, in which the speed of light c plays a fundamental role.

The photon is a really microscopic particle. To get an idea of its size, let us consider a lamp that radiates light with a power of 40 W. The light it emits in 1 billionth of a second contains about 100 billion photons!

The production of an electron-positron pair is a phenomenon understandable only if the light beam is interpreted as a collection of photons, particles in which all the energy needed for the creation of mass is concentrated.

The Nature of Light

Is light a wave or is it a particle? This is a much-debated issue, reflecting its complex and non-intuitive nature. The Danish physicist Niels Bohr (1885–1962) proposed the 'complementarity principle,' according to which the light has a dual nature, depending on the phenomenon considered. It can behave either as a wave or as a particle, but its dual aspect, both undulating and corpuscular, cannot be observed during the same experiment. Is light a wave or a particle? It depends on the phenomena we are interested in! (Fig. 2.23).

Today we know that light is not a wave but is made up of particles called photons. However, these have very peculiar behavior; they are quantum objects to which a probability wave is associated. Their trajectory is not completely determined and may lead along different paths. Scientists talk of probabilistic behavior. Let's see how this property of the photon can explain interference, a phenomenon previously considered unquestionable evidence of the wave nature of light.

Today we are able to create and detect a very faint flow of light, consisting essentially of one photon at a time. If we project this weak beam onto a wall through two slits, what do we expect to appear on the screen?

76 Visible and Invisible

FIGURE 2.23 In 1925, the English scientist Joseph Thomson (1856–1940), discoverer of the electron, exposed the issue of the nature of light in these terms: "The wave-particle view of physics is like a struggle between a tiger (*top*) and a shark (*below*); each is supreme in his own element but helpless in that of the other"

Under these conditions, if a photon passes through a single opening, you cannot get the sum of the rays from the two slits, a combination necessary to achieve interference according the wave model. A photon at a time also appears on the screen, and over

Experiments with Light 77

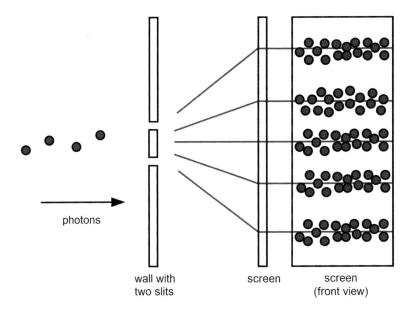

FIGURE 2.24 The probabilistic behavior of photons can be demonstrated by an experiment that envisions the use of a weak light beam and a wall with two slits

time the different traces form an interference pattern, which is therefore not created by the sum of two waves but by the quantum nature of photons, which can travel along different trajectories and then pass through one slit or the other (Fig. 2.24).

Proof of the probabilistic behavior of the microscopic world was provided only recently, in what is considered the most beautiful experiment in physics. The Italians Pier Giorgio Merli (1943–2008), Gian Franco Missiroli (b. 1933), and Giulio Pozzi (b. 1945) were the first, in 1974 (the publication of their results dates to 1976), to experimentally verify the bizarre nature of the microscopic particles, using a very weak flow of particles, containing only one electron at a time.

On YouTube, there are two videos that illustrate the interference of a single electron: 'L'esperimento più bello della fisica l'interferenza di elettroni' (the original experiment, in Italian) and 'Electron Particle Vs. Wave Duality' (in English).

In 2002, from the pages of the magazine *Physics World*, the philosopher of science Robert P. Crease launched a survey to identify the most beautiful experiment in physics. The one that

78 Visible and Invisible

received the most votes was that relative to the interference of single electrons. The top ten also included Galileo's experiment on falling bodies, that of the Millikan oil drop to measure the charge of the electron, and the breakdown of the colors of sunlight with a prism as carried out by Newton.

Irradiation and Black Body

Despite appearances, three bodies as diverse as a heater, a light bulb, and the Sun have certain aspects in common. In the first place, all are powered by a source of energy. The boiler provides hot water for the heater, the electric current makes the lamp filament incandescent, and nuclear reactions power the Sun (Figs. 2.25, 2.26, and 2.27).

Furthermore, in all three cases, the energy continuously supplied to the bodies is transferred to the surrounding environment in the form of electromagnetic waves. Therefore all three radiate energy.

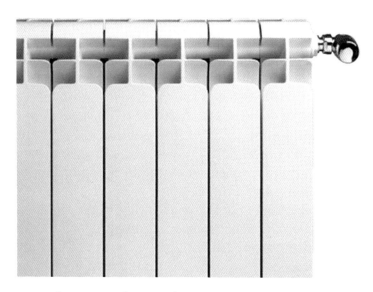

FIGURE 2.25 A heater irradiates infrared rays, invisible to the naked eye

Experiments with Light 79

FIGURE 2.26 A light bulb radiates infrared and visible light

FIGURE 2.27 The Sun emits infrared, visible, and ultraviolet radiation

The amount of energy diffused and the wavelengths of the radiation emitted, however, are different. What are the underlying mechanisms of irradiation?

All bodies are composed of atoms that never stop moving, but constantly oscillate in a motion called thermal agitation, which grows with increasing temperature. These oscillating atoms act as small antennas that continually generate electromagnetic radiation and transfer energy to the environment.

The radiation emitted by a body depends primarily on its temperature. If it is that of the environment, irradiation takes place in the infrared region, with invisible rays detected on the basis of transported heat. This is the case of the heater.

The incandescent lamp, with the filament at 2,400 °C, and the Sun, the outer surface of which is 5,500 °C, both emit radiation in a different region of the electromagnetic spectrum.

In addition to temperature, the type of emission also depends on the shape, surface, and material the body is made up of; therefore, it is difficult to predict the type of radiation diffused. There is, however, a case where such radiation depends only on the temperature, with a well-known universal formula. We are talking about the black body, so-called because it absorbs all the radiation that strikes it.

A black body does not reflect light, but does not necessarily always appear black because of its own radiation. At room temperature it emits infrared rays, but at high temperatures, around 1,200 °C, it gives off white light while continuing to absorb all light striking it.

The radiation emitted by such objects varies as a function of wavelength, giving rise to the spectrum of the black body, which has a characteristic 'bell' shape, with a peak of maximum emission. If the temperature of the body increases, the bell grows rapidly and the peak of maximum emission shifts to lower wavelengths or higher frequencies (Fig. 2.28).

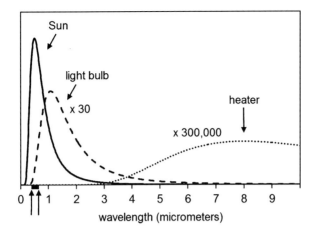

FIGURE 2.28 The graph displays the spectra of a black body at three different temperatures. The *solid line* refers to the external temperature of the Sun (5,500 °C), the *dashed line* to that of the filament of an incandescent bulb (2,400 °C), the *dotted line* to that of a heater (90 °C). The height of the lines, which indicates the intensity of the radiation emitted, is very different in the three cases. The curve at high temperatures significantly outweighs the other two. To display them all together, the height of the *dashed line* was amplified 30 times, while that of the *dotted line* by a factor of 300,000! On the horizontal axis of the graph the *vertical arrows* indicate the part corresponding to visible light, from 0.4–0.7 μm. As we can see, the Sun emits in the ultraviolet, the visible, and the infrared ranges; the incandescent lamp in the visible and in the infrared; and the heater only in the infrared

THE CAVITY OF KIRCHHOFF

Around 1860, the German scientist Gustav Kirchhoff gave the definition of the black body and formulated the general laws that govern it, as well as offering an example of how to create one of these bodies. If we place an object inside a closed cavity equipped with reflective walls, the radiation given off by the matter remains within the volume of the cavity, continuously reflected by the walls. The body, at the same time, emits and absorbs the radiation of the cavity, reaching a state of equilibrium. In these conditions the system (object + cavity) becomes a perfect black body, which absorbs all the waves regardless of their frequency, and the diffusion depends only on temperature. A small opening in the cavity (A) allows us to analyze the inside (Fig. 2.29).

82   Visible and Invisible

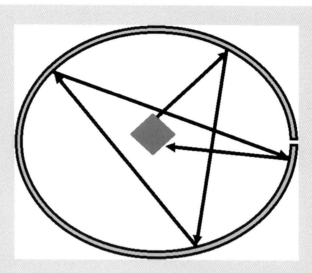

FIGURE 2.29 Schematic representation of the cavity of Kirchhoff

> **CONSIDER THIS**
> Why can't you trap light in a box, even if it is equipped with reflective inner walls?

Transparency

The transparency of a body is dictated by the behavior of the light absorbed by it. Here are two extreme cases:

- *Opaque material.* The non-reflected rays of light are absorbed and do not emerge from the opposite side from that struck (Fig. 2.30).
- *Transparent material.* The light, albeit lessened, passes through the body (Fig. 2.31).

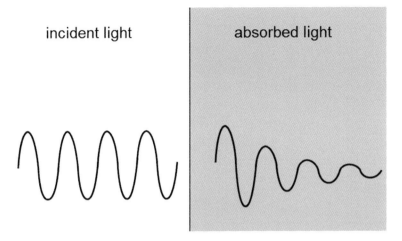

FIGURE 2.30 An opaque body absorbs all the luminous rays that are not reflected

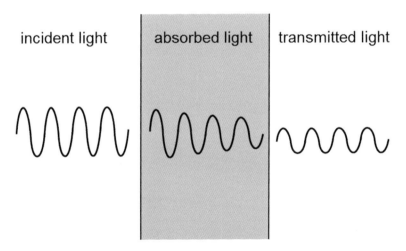

FIGURE 2.31 A transparent body is crossed by light, which is absorbed only partially. Hence, the light that comes out of the body, after crossing it, is less than the incident light

> **VARIABLE TRANSPARENCY**
> The transparency of bodies depends on the wavelength of the radiation striking it. For example, color filters are materials that allow the passage of a single color, absorbing the others. And while our bodies are opaque to visible light, X-rays may pass through them. Even Earth's atmosphere has major properties of transparency, essential for the life forms we know. It is crossed by visible light and radio waves, while at least at ground level, it is opaque to more energetic ultraviolet radiation, X-rays, and gamma rays.

Reflection and Diffusion

What happens when a surface cannot be crossed by the light beam striking it? It reflects it, i.e., each ray of the beam behaves like a sphere (like a tennis ball) bouncing on the floor. We may observe various effects depending on the characteristics of the surface (Fig. 2.32). If it is smooth, we have reflection, i.e., the light beam is reflected as a single shaft, since all its components undergo parallel reflections (Fig. 2.33). On the other hand, if the surface is rough, the light is diffused in all directions, because each component is reflected in a different direction from the others.

Therefore, when a smooth surface is illuminated, the rays reflected can be seen only in a certain position, that mirroring the direction in which it was struck. If the surface struck is not reflective, the light is sent in all directions. Thus the image is diffused in space, and each observer, wherever he is, may receive a ray of light from the object in any location.

Mirrors

Every smooth and reflective surface is a mirror, producing an image of the objects it reflects.

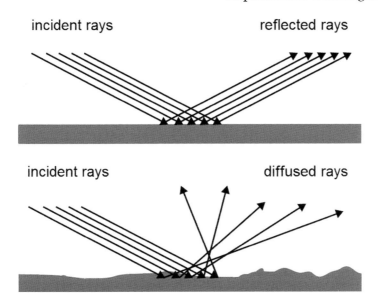

FIGURE 2.32 The rays incident on a smooth surface are reflected. Those incident on a rough surface are diffused

FIGURE 2.33 It is sometimes difficult to distinguish a picture from its image inverted by reflection. For example, which of the figures above is the right way up?

How does this mechanism work? Let's think about what happens when a body is placed in front of a mirror. Each ray that strikes it is reflected in it symmetrically. Let's consider some rays, for example, those from the dot on the spinning top in Fig. 2.34.

86 Visible and Invisible

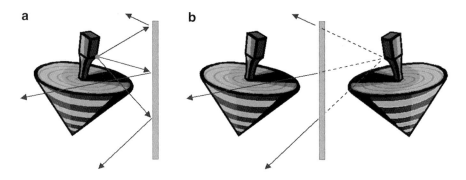

FIGURE 2.34 A spinning top. The image (**a**) formed by the reflected rays is the same as would occur without the mirror, with the object placed symmetrically behind the reflective surface (**b**). The luminous rays in fact seem to come from behind the mirror, even though in fact none of them cross it; for this reason we speak of a virtual image

CONSIDER THIS
When we look at ourselves in a mirror, we clearly see our own image, but we do not see ourselves as others see us. Why? (Fig. 2.35).

FIGURE 2.35 In the mirror we do not see ourselves as others see us. This German stamp from 1962 depicts the Queen, Snow White's evil stepmother, gazing into her looking glass

Refraction

Light can pass through various substances, for example, from air to water. When it crosses the interface between the two, a portion of the beam is reflected while the rest enters the new medium.

Light rays that enter a dense body such as water or glass, called reflected rays, display a remarkable feature. They do not continue straight but are deflected. Even the reflected beams change direction, but their bouncing is symmetrical. The reflection angle, between the direction of the reflected rays and the one normal to the surface separating the two media, is equal to that of incidence, formed from the same normal and from the direction of the incident beam. The angle of refraction, on the other hand, formed between the direction of the refracted rays and the normal, is different (Fig. 2.36). Why does this happen? When light passes through a dense medium, its velocity decreases, as well as its wavelength. Accordingly, the beams tilt, except when traveling perpendicularly to the surface of separation. In this case they proceed along their initial trajectory.

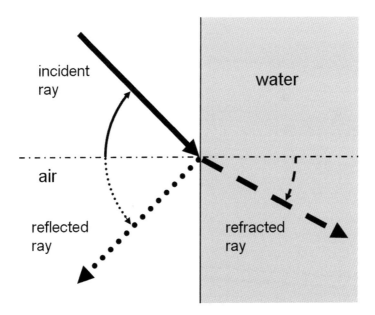

FIGURE 2.36 When a light beam goes from air to water, the angle of reflection (*dotted*) is equal to that of incidence (*solid*), while that of refraction (*dashed*) is different

88 Visible and Invisible

Refraction takes place not only with light; this phenomenon occurs with all waves—electromagnetic, acoustic, and those in a liquid.

> **CONSIDER THIS**
> From the shore of a lake or from a beach it is sometimes possible to observe fish swimming. As a result of refraction, the fish seem to be closer to the water surface than they actually are. Why is that? (Fig. 2.37)

FIGURE 2.37 The path of the beam that transversely strikes the surface separating two substances deviates. This causes various optical phenomena, such as the apparent bending of partially submerged objects or the apparent displacement of the image as it passes through a liquid

Prisms

When light crosses the interface between two different media, it is deflected, and each ray changes direction differently depending on its color. Red is deflected less than violet, and the deviation of other colors is in the middle between these two extremes.

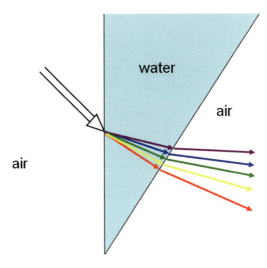

FIGURE 2.38 If a light beam passes through two non-parallel air/water and water/air surfaces, it is broken down and produces a rainbow both in the water and air

In a dense medium such as water, colors are separated by refraction and form a rainbow. But what happens when light passes through water and then returns to pass through the air?

If the second separation surface is parallel to the first, each color component resumes its initial direction. For example, the rays re-form into white light, and the beam that initially struck the surface is simply shifted. In this case, the rainbow disappears.

To preserve the rainbow even after the crossing water/air, it is necessary for the two separation surfaces not to be parallel, as shown in Figs. 2.38 and 2.39.

In conclusion, when light is initially in one medium and passes through a second before returning to the first, the refraction separates the various colors only in the presence of these two conditions:

- the incident beam is tilted compared to the first surface of separation;
- the two surfaces of separation are not parallel.

These conditions are fulfilled if we use a triangular glass prism. In this case, if a light beam (tilted compared to the first surface) strikes the prism, refraction first takes place inside the glass and again in the moment it comes out the other side and passes through the air once more (Fig. 2.40).

90 Visible and Invisible

FIGURE 2.39 The same thing as in Fig. 2.38 may be achieved by using a glass prism

FIGURE 2.40 Isaac Newton, portrayed by Sir Godfrey Kneller in 1702. The great English physicist began to deal with the luminous phenomena around 1660, going on to formulate the theory of colors among other things

Experiments with Light

CONSIDER THIS

In 1973, Pink Floyd released the album *The Dark Side of the Moon* (issued by EMI), the cover of which showed the image of a ray of light refracted by a prism, as shown in Fig. 2.41, thereby committing an error. What was it?

FIGURE 2.41 The image on the cover of the Pink Floyd album

Lenses

A lens, glass, or plastic is the simplest of optical instruments. The first refraction occurs when light enters, the second when it comes out. Because the surfaces that form it are not parallel, the light beam that passes through it is deflected (Fig. 2.42).

There are two types of lenses: converging and diverging. In the case of the former, the central region is thicker than the edges, and an incident beam (parallel to the axis of the lens) converges at a point known as the focal point or focus (Fig. 2.43). In the latter case, the central area is thinner and the rays of the beam are refracted so as to diverge. Below, we shall examine a converging lens.

Each lens has two foci (focal points), located on opposite sides of the lens itself. If an object is placed at a focal point and then viewed from the other side of the lens, it appears magnified.

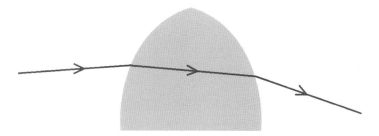

FIGURE 2.42 The deviation of a light beam by a lens

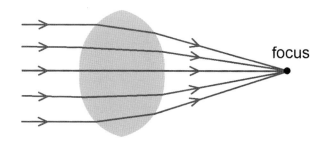

FIGURE 2.43 A beam of parallel rays striking a (converging) lens 'converges' at its focal point

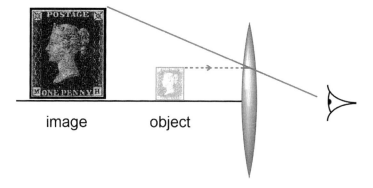

FIGURE 2.44 From every area of an object placed at the focal point, a single beam reaches our eye, deflected by the lens (*dotted line*). Since our eye interprets the signals as straight (*solid line*), we perceive the image as enlarged

Let us see why. Each detail of an illuminated object gives off light in all directions. From each point, a single beam reaches the observer, in this case through a curved path from the lens, according to the scheme shown in Fig. 2.44, in which the beam emitted

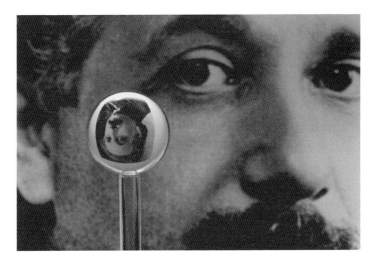

FIGURE 2.45 Example of overturning due to the distance between the lens and the object observed

by the object is indicated by the first part of the dotted line, which then continues through the lens and to the eye.

The eye, like all optical devices, always interprets the signals received as generated by straight rays as if they had not undergone deviation. As far as our visual perception is concerned, the light comes from an expanded image and is the result of the magnification of the object (Fig. 2.44).

If the body observed through the lens is far away, the rays that strike it travel parallel to one another. In this case the image is formed in the focal point but reversed (Figs. 2.45 and 2.46)! This overturning normally happens in a dark room, in a camera, and in the eye.

Often lenses are not used individually but combined. For example, for a microscope or a telescope, at least two are required.

The fundamental material for the construction of good lenses is glass, a transparent and rigid compound, capable of bending light rays and made of oxygen and silicon, atoms that are also found in silica sand and quartz. Pure quartz glass melts at high temperatures, above 1,700 °C, but the addition of substances such as soda (sodium carbonate) lowers the melting point, making it easier to process.

Who discovered glass? The Phoenicians were masters of the art of glassmaking, around the seventh to sixth century B.C., but

94 Visible and Invisible

FIGURE 2.46 When an object is placed in the focus of a lens and is illuminated by a powerful source, such as the Sun, the temperature of the object grows rapidly and can even catch fire

the discovery of glass was certainly earlier, probably dating back to Mesopotamia in the third millennium B.C.

The Nimrud lens is a piece of pure quartz, 3,000 years old, unearthed during excavations in the palace at Nimrud in northern Iraq. Assyrian craftsmen, who created complex carvings, might have used it as a magnification tool in their works.

In *Naturalis Historia* (Natural History) written by Pliny the Elder (A.D. 23–79), the following legend is told.

> *In Syria there is a region known as Phoenice, adjoining to Judaea, and enclosing, between the lower ridges of Mount Carmelus, a marshy district known by the name of Cendebia. In this district, it is supposed, rises the river Belus, which, after a course of five miles, empties itself into the sea. The story is that a ship, laden with niter, being moored upon this spot, the merchants, while preparing their repast upon the seashore, finding no stones at hand for supporting their cauldrons, employed for the purpose some lumps of niter which they had taken from the vessel. Upon its being subjected to the action of the fire, in combination with the sand of the seashore, they beheld transparent streams flowing forth of a liquid hitherto unknown. This, it is said, was the origin of glass.*

Interference

One of the characteristic phenomena of light is interference, due to the superposition of several waves that have traveled different paths in a single point of space. It is a phenomenon easily understandable within a wave framework, because the waves show peaks and troughs that, when superimposed, are able to generate different effects. If one adds two maxima (or two minima), the signal increases, while if a maximum is superimposed onto a minimum, the signal is reduced.

The resulting wave may cancel itself, so that it does not result in any visible effect, or, conversely, reaches an intensity equal to the sum of the intensity of the overlapping waves. If the resulting wave has an intensity greater than that of each wave component, we speak of constructive interference; otherwise it is known as destructive interference.

An effective tool for creating interference is the grating, i.e., a system consisting of a set of holes in an opaque slab (Fig. 2.47). The light striking such a plate can be split into different paths passing through the openings, and the interference takes place when the wavelength is close to the distance between the holes. In the case of visible light, the pitch of the grating must be in the order of the micrometer. Under these conditions, we speak of an optical grating. The instrument can produce various effects, including the formation of a rainbow.

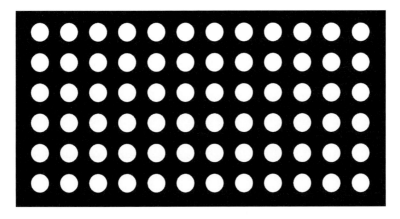

FIGURE 2.47 A grating made up of a set of holes in an opaque slab

96 Visible and Invisible

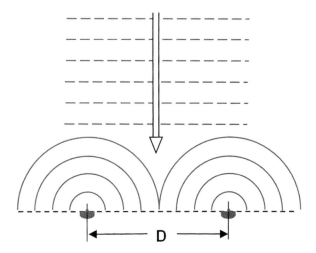

FIGURE 2.48 The diffusion of incident light generated by a small area of a CD, represented as a flat surface (*dashed line*) and two grooves. The reflection is described in terms of wave fronts, *continuous lines* obtained by joining the maxima of the waves. The fronts are parallel due to the rays reflected by the surface, and circular due to those caused by the grooves

There is not only the transmission grating, formed by a set of openings. It is possible to obtain the interference also with a reflection grating, in which an ordered system of grooves reflects the incident rays.

As an example, let us consider the surface of a compact disc. If we look at a small area of a CD, with only two grooves, separated by distance D, we may suppose that the light strikes the surface perpendicularly (Fig. 2.48). Each groove diffuses light in all directions, while the flat and mirroring areas reflect it. Observing the surface from an inclined direction, as in Figs. 2.49 and 2.50, we receive the sum of the rays from the various engravings: this sum produces the interference.

At some points, such as point A in Fig. 2.49, the maximum of the wave coming from the groove on the left is combined with the minimum of the wave coming from the one on the right, and the same goes for thousands of other pairs of engravings on the CD surface. As a final result, an observer at A receives a weak signal, i.e., point A is poorly lit. We are thus in the presence of destructive interference.

In other points, such as B in Fig. 2.50, the maximum of the wave coming from the left side of the groove is combined with the

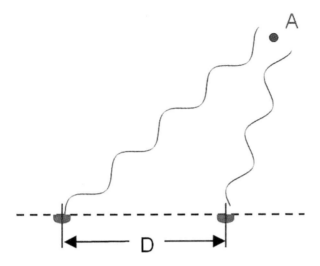

FIGURE 2.49 Example of destructive interference. At point *A* the maximum of the wave on the *left* and the minimum of that on the *right* are summed

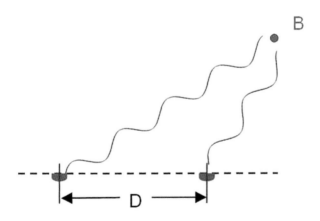

FIGURE 2.50 Example of constructive interference. At point *B* the maxima of the wave on the *left* and of that on the *right* are summed

maximum of the wave from the right-hand one, and the same goes for the other thousands of pairs of engravings on the CD surface. As a final result, an observer at B receives a strong signal, that is, point B is very bright. We are thus in the presence of constructive interference.

Up to this point, we have considered the effect of monochromatic light with a single wavelength. In this case, interference simply produces bright areas, interspersed with less luminous sections. What happens if instead we use white light, which contains several wavelengths, each corresponding to a color?

In this second case, the sum of the waves depends on the wavelength of the individual rays and produces different points of minimum and maximum for the various colors. Consequently, such interference is no longer visible. On the other hand, for each color we have a distinct point of maximum illumination. The grating is thus able to separate the colors of white light. This is the reason why a CD is able to split the various colors and form a rainbow. It is in fact made up of thousands of tracks spaced 1.6 µm apart (Fig. 2.51). The colors are deflected at different angles, according to the Table 2.4 below (Fig. 2.52).

TABLE 2.4 Light deflection by a CD

Wavelength (µm)	Deflection angle (degrees)
0.400	14.5
0.450	16.3
0.500	18.5
0.550	20
0.575	21
0.600	22
0.650	24

As seen in the table above, the angle of deflection of a light ray varies according to its wavelength. The shorter the wavelength, the lesser the deflection

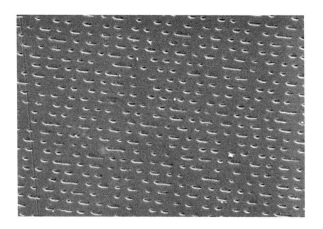

FIGURE 2.51 An example of an optical grating. Here is what a CD zoomed 2,000 times with an electron microscope looks like

Experiments with Light 99

FIGURE 2.52 Due to the phenomenon of interference, a ray of light striking a CD generates a rainbow on the surface

3. Light and the Sky

"I have no special talents. I am only passionately curious."

Albert Einstein, scientist and Nobel laureate

The Color of the Sky

The white light of the Sun is a mixture of all colors. In a vacuum, it travels in a straight line and illuminates only the objects in its path; that's why astronauts see the sky as black, even in daylight.

Before reaching us, solar rays pass through the atmosphere, the gaseous layer that surrounds Earth. In this region, light illuminates space due to a phenomenon called diffusion (Fig. 3.1).

The spectral colors have different weights in diffused light, where blue and violet prevail (Figs. 3.2 and 3.3). The bright sky is therefore dominated by these hues, but we see it as blue because our eyes have very little sensitivity to violet. The enrichment of the colors blue and violet in diffused light leads to the depletion of these hues in transmitted rays, which instead are dominated by red and orange (Fig. 3.4).

CONSIDER THIS
Why is the Sun red at sunrise and sunset?

CONSIDER THIS
Water has an intrinsic color, which can be observed when light passes through it at a thickness of several meters. This hue is bluish and is due to the same phenomenon that determines the color of the sky. Can that be true?

102   Visible and Invisible

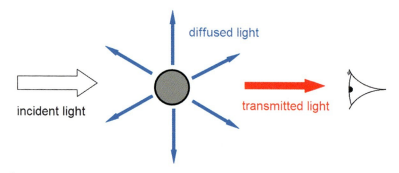

FIGURE 3.1 Earth's atmosphere is made up of molecules that can interact with light. When these are hit by a light beam, it absorbs some of it, while the remainder continues as transmitted light. The radiation absorbed momentarily excites the individual molecules; these then return to their initial state by emitting light in all directions, shooting off like little antennas. Thanks to this diffusion effect, the sky appears clear to us

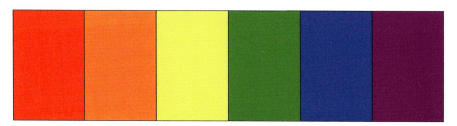

FIGURE 3.2 Incident light. No color dominates

FIGURE 3.3 Diffused light. *Blue* and *violet* dominate

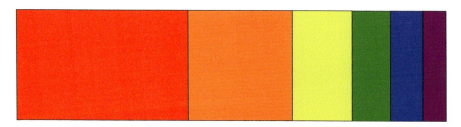

FIGURE 3.4 Transmitted light. *Red* and *orange* dominate

Daylight

The movement of Earth around the Sun produces the cycle of the seasons, while the rotation around its own axis generates the succession of day and night. Throughout the progress of the day, the position of the Sun as observed from Earth varies continuously, generating a trajectory described as ecliptic, shown by the dashed curve in Figs. 3.5 and 3.6, at which the highest peak reached every day is at noon.

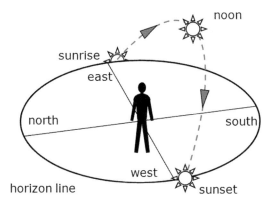

FIGURE 3.5 The ecliptic, the Sun's apparent trajectory in its daily motion

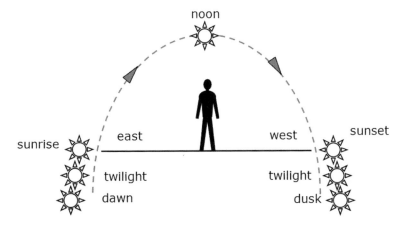

FIGURE 3.6 The Sun rises in the east and sets in the west, reaching the maximum height along the ecliptic at noon. Before sunrise we can experience the phenomena of dawn and morning twilight, after sunset that of the evening twilight and dusk

The sky is already bright when the sun rises above the horizon to the east, while at sunset, when it disappears to the west, the light is slow to fade.

Even before the Sun has risen, before its direct light reaches us, the sky above us is already illuminated. This first flare denotes the dawn. Shortly before the appearance of our star there appears a diffused light of various colors, from yellow to violet. This is the morning twilight (Fig. 3.7).

At dawn, the molecules of the atmosphere spread light in all directions, illuminating regions that would otherwise be dark, and the dominant color is light blue. The duration of this effect varies with latitude. It is minimal at the equator (lasts for only a few minutes) and at its longest at the poles (for a few hours). In New York City, it lasts about 30 min.

Similarly, at sunset, when the Sun disappears, the atmosphere continues to diffuse its light, giving rise to evening twilight and dusk (Fig. 3.8). Due to the unusual and romantic quality of the lighting, twilight is renowned among photographers and painters, who also call it the 'blue hour' or 'sweet light.'

Like at dawn, also at dusk the duration is longest in the Arctic and Antarctic regions. Twilight can brighten the sky during the polar nights, even for several months on end.

CONSIDER THIS
Why does the Sun rise earlier and set later in the summer, making the day last longer than the night?

Rainbows

When a rainbow is formed, the various hues that are normally mixed together in white light are separated and form one or more colored arches. The main one, the primary rainbow, features an inner part colored violet and the outermost one colored red (Fig. 3.9).

Color separation occurs when the luminous rays pass through transparent bodies with particular forms. There are in fact various kinds of geometry capable of splitting up hues. The form most

Light and the Sky 105

FIGURE 3.7 Morning twilight features diffused light ranging from *yellow* to *violet*, which appears at dawn just before sunrise

FIGURE 3.8 In the evening, after the Sun has disappeared over the horizon, we see an analogous phenomenon, in which the color tends to *blue*. This is the evening twilight

106   Visible and Invisible

FIGURE 3.9 In a primary rainbow, we see red in the outermost zone and violet in the innermost one

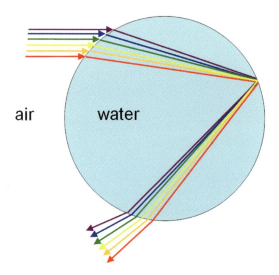

FIGURE 3.10 When white light enters a water droplet, the spectral colors are separated by refraction, then they are reflected and finally, when they emerge from the droplet itself, are subject to secondary refraction

commonly found in laboratories is that of the prism, but also a sphere can separate the hues. This is the case of the rainbow, created by the spherical drops of water in the air during the rain.

Figure 3.10 shows a possible path of the various colored rays that make up white light when they encounter a water droplet.

Entering the sphere they undergo initial refraction; they are then reflected when they hit the inner surface, and lastly they exit, undergoing secondary refraction. At the end of this process, the rays come back along a new path. Since the refraction separates the colors, the final deviation is different for each component of the visible spectrum. Red light is deviated by 42°, while violet by 40°, and the other colors fall between these two extremes.

The hues thus separated from the water droplets travel in different directions, forming the rainbow. To see this effect it is necessary to have the Sun behind you and find yourself in a position where it is possible to intercept the rays deviated by 40°–42° compared to the direction of incidence (Fig. 3.11).

There are six hues, called spectral colors, of the rainbow: red, orange, yellow, green, blue, and violet.

> **THE MOON RAINBOW**
> Aristotle, in his treatise *Meteorology*, refers to the night rainbow, created by the Moon, in these terms: "We have only met with two instances of a moon rainbow in more than 50 years (Fig. 3.12)."

> **CONSIDER THIS**
> Aristotle argued that the rainbow has the shape of a circular arc that cannot be larger than a semicircle. Is he right?

> **CONSIDER THIS**
> Sometimes you can see a second rainbow, separated from the first, less bright and with the colors inverted, and the sky also appears darker between the two rainbows. In this case, the phenomenon is due to refraction and reflection. Why is this rainbow so different from the primary one?

108 Visible and Invisible

FIGURE 3.11 In order to observe a rainbow one must be in a particular position

FIGURE 3.12 An example of a Moon rainbow, in which the light of our satellite is deflected by the drops of water vapor produced by Victoria Falls. The constellation of Orion is visible behind the upper part of the rainbow

Glories

It happens occasionally that the water droplets that form a cloud send back part of the sunlight that strikes them with mechanisms other than those that give rise to the rainbow. In these exceptional cases, a person standing between the Sun and the cloud may glimpse his or her own shadow, hugely magnified, projected onto the cloud. Furthermore, the reflected light is often surrounded by different colored rings. This phenomenon, now fully understood by science and called a glory, has long been a mysterious event that very much impressed the ancients (Figs. 3.13 and 3.14).

The main difference between a rainbow and a glory lies in the rebound mechanism caused by the droplets. In the rainbow, light undergoes two refractions and one reflection, with a final deviation of 40–42°; in the glory, on the other hand, the outgoing rays move in the same direction as the incident ones, only they travel backwards.

How can it be that water droplets send back light?

When a beam of light hits a drop of moisture, its rays may follow very different paths. Some are reflected from the surface, while others enter the drop. Of the latter, some pass through it to then go out into the air; others may be reflected from the inner surface, bouncing once, twice, or even more times.

FIGURE 3.13 To see a glory, the observer must stand between the Sun and a cloud

110 Visible and Invisible

FIGURE 3.14 Since the phenomenon is visible in the opposite direction from the Sun, today it is observed most easily when you are in flight, with the glory surrounding the shadow cast by the airplane

Some incident rays, finally, may hit the edges of the droplet, following the path shown in Fig. 3.15. They are reflected, bouncing around the sphere, and then complete their trajectory on the basis of a new effect—the surface wave.

The light that travels in this way is coupled to the surface of the droplet and follows its shape, finally exiting the drop in the direction parallel to its direction of entry, but traveling in the opposite direction.

The process described above, however, is but one of the possible paths for the luminous rays that hit a drop to follow. The light deflected in this way is minimal, and the signal generated is weak. So just how is a glory formed?

This effect is caused by the wave nature of light. Let's see how.

- All points of the edge of a given droplet send the same signal, and these rays travel the same distance to reach the eye, with a synchronized maximum and minimum. The superposition of

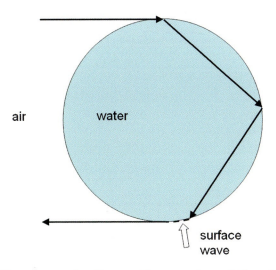

FIGURE 3.15 The process leading to the creation of the glory, with the surface wave which ends its path

so many waves of this kind produces constructive interference for all frequencies (all colors), creating a bright white beam in the center.
- The waves originating from the drops close to those considered do not have the same symmetry. The rays from the edges run along slightly different distances, and the various wavelengths cancel one another out—destructive interference. Under these conditions, the luminous signal returned is minimal.
- If we consider drops even more distant from the center, we may have constructive interference when the difference between the paths of the various waves is equal to a wavelength. This condition occurs first with violet (the shortest wavelength), then with the other hues. The effect therefore creates colored rings around the central light white, with a red outer band.

112 Visible and Invisible

🔍 GLORIES IN LITERATURE

In the mid-1940s, the American writer Henry Miller (1891–1980) went to live in Big Sur, a charming place on the central coast of California, where the Saint Lucia mountains stand out against the Pacific Ocean. In this place Orson Welles and his wife, Rita Hayworth, bought a log cabin, which is now owned a restaurant (the Nepenthe). Here is what Miller said in his description of the glory, viewed from the mountains of Big Sur:

"If it be shortly after sunup of a morning when the fog has obliterated the highway below, I am then rewarded with a spectacle rare to witness. Looking up the coast toward Nepenthe, where I first stayed (then only a log cabin), the sun rising behind me throws an enlarged shadow of me into the iridescent fog below. I lift my arms as in prayer, achieving a wingspan no god ever possessed, and there in the drifting fog a nimbus floats about my head, a radiant nimbus such as the Buddha himself might proudly wear (Fig. 3.16)."

FIGURE 3.16 The shadow of a man, projected onto a cloud and surrounded by a glory

GLORIES IN RELIGIONS

The colored glory around the shadow of an individual was considered an indication of personal enlightenment, often associated with the divine and used as a symbol of holiness in sacred art, by Christian as well as other artists (Fig. 3.17).

FIGURE 3.17 *Christ among the Apostles, a detail from* The Tribute Money, *a famous fresco by Masaccio*

Aurorae

The Sun not only sends out electromagnetic waves but also the solar wind to Earth, consisting of microscopic electrically charged particles, mainly protons and electrons (Fig. 3.18). This life-threatening flow does not reach us thanks to a shield created by Earth's magnetic field, the same that orients our compass.

114   Visible and Invisible

FIGURE 3.18 Image of the solar wind and of the activity of the Sun processed by the SOHO satellite, as part of a European Space Agency (ESA) and NASA mission. The photo shows the Sun in ultraviolet light, in a region that extends over 2 million km from the solar surface

Earth's magnetism extends into space for tens of thousands of kilometers. The solar wind confines the Earth's magnetic field in a region shaped like a comet: the magnetosphere.

When the solar wind particles reach the upper atmosphere, where the magnetic field is intense, they are diverted and glide along the outer edge of the magnetosphere. Sometimes, however, the charged particles can penetrate Earth's atmosphere, creating aurorae. Let's see how this can happen.

The solar wind not only carries particles but pulls well beyond Earth's orbit the Sun's magnetic field, giving rise to the interplanetary magnetic field, which leads to continuous variations related to solar activity and which is opposed to Earth's own terrestrial field (Fig. 3.19). The strongest effects are observed during periods of intense solar activity, when the interplanetary field weakens Earth's magnetic shield, creating channels through which a large amount of protons and electrons flow.

The protons (with their positive charge) are turned away from Earth, while the electrons (negative charge) are pushed down into

FIGURE 3.19 In the drawing, the solar wind is shown with *clear lines*; Earth is the small sphere on the *right* and is surrounded by the wide terrestrial magnetic field, shown by *violet* and *blue lines*

the ionosphere. The aurora is formed by the electrons that strike the atoms of the upper atmosphere of Earth. In the collision, the former transfer energy to the latter, which return it in the form of light (Figs. 3.20 and 3.21).

Most of the auroral phenomena take place at the Arctic regions, which have less magnetic protection. Aurorae near the North Pole are called the aurora borealis, or the northern lights, while those near to South Pole are called the aurora Australis, or the southern lights.

> **FOX FIRES**
> The aurora is a wonderful sight, consisting of vapor light, dancing, and swirling. Ancient peoples often associated fantastic meanings with and explanations for these events. The Greeks described them as a rain of blood, the Inuit attributed the cause to the inhabitants of heaven, the Vikings thought they originated from the sunlight reflected by the shields of the Valkyries, while the ancient inhabitants of Lapland believed that the aurora was caused by a huge fox that, with his tail, would sweep the celestial vault. Still today, the Finnish word to indicate the aurora is *revontulet*, or literally, 'fox fires.'

GEOMAGNETIC STORMS

The instability of Earth's magnetosphere, caused by solar activity and by the currents that penetrate the atmosphere, can give rise to electromagnetic storms that cause problems with electricity distribution networks, or interfere with the operation of radio communications.

In March 1989, astronomers observed a major solar flare. A few days later, an enormous perturbation in the solar wind reached Earth and the aurora borealis, seen even in Florida, Cuba, and Mexico, was gigantic. A powerful stream of electric charge began to spin at an altitude of 100 km above North America, inducing electrical currents in the terrestrial soil that soon found the ideal circuit in which to flow—power lines. On March 13 at 3 a.m., the collapse began. In less than a minute 50 % of the electrical networks of Québec shut down. Millions of people were affected. In Toronto the temperature that night was –6.8 °C; the blackout lasted 9 h, and the total damage was estimated at $3 to $6 billion.

FIGURE 3.20 Aurora borealis above Bear Lake in Alaska

FIGURE 3.21 Aurora australis, photographed in 2010 by the International Space Station (ISS)

The United States was also affected, with damage to more than 200 transformers and repeaters. A collapse like that in Canada was narrowly avoided thanks to a dozen 'heroic' capacitors in the Pennsylvania power line that continued to operate, despite the major voltage fluctuations.

The Green Ray

The green ray is a very rare and fascinating phenomenon, visible only for 1–2 s at sunset. The upper part of the solar disk suddenly changes color from yellow–orange to a wonderful green of spectral purity, which cannot be conveyed adequately by a photographic image; only seeing it first-hand allows you to fully appreciate it.

How can we witness this rare event? It is best to observe the sunset on an unhindered horizon, such as that of the sea (Fig. 3.22).

The green ray is nothing more than a sunset special, even though all sunsets are something special.

118   Visible and Invisible

FIGURE 3.22 The phenomenon of the *green ray* observed at sea

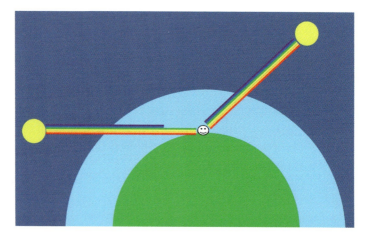

FIGURE 3.23 During daylight hours the sunlight appears *white* to us; it turns *red* at sunset because the Sun's rays follow different paths at those times—a shorter one during the day, a longer one at sunset

In Fig. 3.23, the yellow spheres represent the Sun at two different times of the day, during daylight hours (right) and at sunset (left); the green sphere is Earth, and its atmosphere is colored blue.

During daylight hours, all visible sunlight can pass through Earth's atmosphere and reach us as white light. When the Sun is setting, however, it appears red to us because its light, traveling through a longer stretch of atmosphere, loses its violet and blue

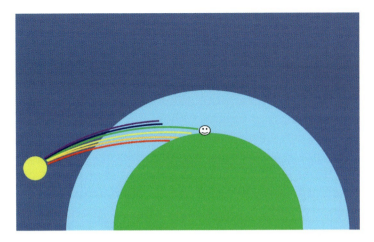

FIGURE 3.24 The atmosphere causes a deflection of the light rays, smaller for *red* and greater for the *violet*. In the drawing, the curvature of the rays has been greatly exaggerated

colors, which are then fully diffused. For the sake of simplicity, in the image the Sun's rays are straight, not bent by refraction.

Furthermore, the atmosphere refracts light continuously, deflecting it more and more as the air becomes denser. For this reason, when the Sun appears to be on the horizon at sunset, it has actually already eclipsed (Fig. 3.24).

The bending of the rays is not the same for all colors. Red is deflected less than blue. Consequently Earth hides from view before the rays with less curvature, such as red, orange, and yellow. For 1 or 2 s they disappear, leaving the green, blue, and violet rays in the sky. Since violet and blue are dispersed by the atmosphere, it is possible to glimpse the green as the last remnant of light.

THE GREEN RAY FROM VERNE TO ROHMER

In 1882, the writer Jules Verne wrote a novel entitled *The Green Ray*, a phenomenon described in these terms: "It will be green, but a most wonderful green, a green which no artist could ever obtain on his palette, a green which neither the varied tints of vegetation nor the shades of the most limpid sea could ever produce the like!" About 100 years later, Verne's text inspired the director Eric Rohmer to make the film *The Green Ray*, awarded with the Golden Lion at the 43rd Venice Film Festival.

Lightning

Under normal conditions Earth's atmosphere behaves as an insulator that prevents the circulation of electric currents. During thunderstorms, however, the clouds can become electrically charged, accumulating energy. When this is high enough, a particular process can start from within the cloud, during which electrons are stripped from atoms, creating channels of ionized air that act as electrical conductors. It is believed that this is caused by very fast electrons, moving at speeds approaching that of light. The conductive air channel propagates jaggedly, and when it reaches the ground, it causes an electric shock, with a flash of visible light—lightning (Fig. 3.25).

The discharge is fast-moving, heating the air up to 30,000 °C, five times the surface temperature of the Sun. The electric current transported, several thousand times greater than that which

FIGURE 3.25 By examining high-speed film footage, it was discovered that lightning is often made up of more than one discharge, typically three or four. According to an estimate made through the use of satellites, an average of four million lightning strikes are discharged on our planet every day

normally flows in our houses, can destroy any object in its path. The rapid expansion of the air created by the discharge produces an acoustic shock wave—thunder.

The luminous flash travels at the speed of light, 300,000 km/s, while the roll of thunder spreads at the speed of sound, almost 900,000 times slower, so the flash reaches us before the noise, and from the delay between the two we can even calculate the distance that separates us from the lightning. Since it is difficult to hear thunder from a distance greater than 20 km, this calculation can only be done using lightning close enough, in which case the light takes only millionths of a second to reach our eyes.

Consequently, the flash may be considered instantaneous, and the delay of thunder can be associated with the motion of the sound wave, which takes 3 s to cover 1 km. It is therefore enough to divide the seconds of delay by 3 to get the distance, in kilometers, between us and the lightning. For example, if we hear thunder 15 s after seeing the flash, the lightning will be about 5 km away.

What kind of radiation is generated by lightning? In addition to the classic visible glow, the discharge emits in the range of radio frequencies, and this produces the familiar crackle of radio transmissions during a thunderstorm.

Recently, X-ray emission has also been discovered, difficult to detect because it does not cover long distances in the atmosphere. The energy of the X-rays emitted by a lightning bolt corresponds to that of two chest radiographs.

How would lightning appear to us if we were equipped with X-ray vision like Superman? We would see a rapid series of bright flashes descending from the clouds, more and more intense, ending with a strong final explosion at the moment when a dazzling light appears.

Even the most energetic radiation, typically emitted by cosmic cataclysms such as supernovae, is created by lightning. Satellites have detected the presence of gamma rays, known as terrestrial gamma rays, emitted during lightning. This type of radiation can be detected only from the space, because the rays directed towards Earth are absorbed by the atmosphere.

> **SELF-DEFENCE AGAINST LIGHTNING**
> Lightning can be very dangerous, and people have always tried to prevent the damage it can cause. While the ancients tried to protect themselves by offering sacrifices to the gods, thinking that lightning represented their anger, scientists have begun to understand its nature, and therefore how to design prevention mechanisms.

They now know, in particular, the importance of sharp objects that can guide the path of a lightning strike to ground. This discovery was the foundation of the lightning conductor, consisting of a long and thin metal rod placed on top of a building. The discharge is attracted by the tip, which allows the channeling of the electrical current to ground without causing damage to the building. It is thus not a question of avoiding lightning but of rendering it 'harmless,' channeling its energy, and today, even trying to use it.

Lightning is attracted to pointed shapes but cannot penetrate an electrical conductor. For this reason during a thunderstorm it is dangerous to find yourself near a tree or standing on a flat stretch of land like a beach, while it is safe to be in your home or car with the windows closed.

Thunder and lightning often arouse a sense of fear. During thunderstorms, some individuals (especially children) suffer from severe fear and anxiety attacks. This disorder is known as astraphobia and is also reported in dogs and cats.

Earth's Atmosphere

Earth has an envelope of gas, the atmosphere, that separates its surface from the inhospitable environment of space, and which blocks a fair percentage of solar radiation. The gases are retained by gravitational attraction; hence its density decreases with height, and there is no clear separation with the cosmic void. At various altitudes, the properties of the atmosphere change, creating five different layers.

The troposphere extends up to about 10 km in height. This is the layer in direct contact with the planet, which hosts the greatest amount of air. The heating of this zone is mainly due to terrestrial irradiation; for this reason the temperature decreases with

FIGURE 3.26 A shooting star, originating from friction between the mesosphere and the passage of a meteorite

height. Most meteorological phenomena occurs in this region, which also houses most life. All plants and animals live here.

The stratosphere extends between approximately 10 and 50 km in altitude. In this layer the heat comes mainly from the Sun, so the temperature tends to rise with height. The stratosphere, in the range of 20–30 km, is rich in ozone, a particularly reactive kind of oxygen with a molecular structure that absorbs ultraviolet radiation, dangerous to life.

The layer between 50 and 85 km constitutes the mesosphere.

In this region falling stars are generated, i.e., small meteors that burn out due to friction, leaving luminous trails behind them. In the upper part, the air is so rarefied as to provide no palpable resistance to the motion of bodies. For this reason, in astronautics the mesosphere is considered the boundary with space (Fig. 3.26).

The thermosphere begins at an altitude of 85 km and extends up to 690 km. This is the place where aurorae occur and includes most of the ionosphere, which is of great importance in telecommunications. In this region, gas is ionized by the radiation of the Sun, and the electrons thus freed reflect radio waves, allowing them to reach places hidden by Earth's curvature.

The exosphere constitutes the outermost layer of Earth's atmosphere and extends over 690 km altitude. This zone, which ends when the density of its gas is equal to that of interstellar space, at about 2,000 to 2,500 km above Earth's surface, contains the magnetosphere, where Earth's magnetic field captures and deflects the solar wind, thus protecting all living beings.

The complex composition of Earth's atmosphere provides major elements of opacity and transparency, both essential to life. It is transparent to visible light and radio waves, while it is opaque to electromagnetic radiation dangerous to living beings, such as gamma rays, X-rays, and ultraviolet light (Figs. 3.27 and 3.28).

The Color of Planet Earth

Earth is one of only two planets in the Solar System that appears blue (the other is Neptune). Its beautiful color is caused by the presence of water, for oceans, seas, lakes, rivers, and marshes occupy about 71 % of Earth's surface (Figs. 3.29 and 3.30). But what color is water?

If we consider a small amount of it, such as that contained in a bowl, it appears perfectly clear, colorless, and transparent. However, when its mass increases, the color tends towards blue, due to two mechanisms:

- Light diffusion by the water molecules (the same effect that colors the sky);
- The selective absorption of light. Water is not perfectly transparent, and, in large quantities completely absorbs light rays, so darkness dominates the sea, at a depth of about 200 m or deeper. However, the absorption rate is not the same for all wavelengths. Red light is absorbed more readily than blue.

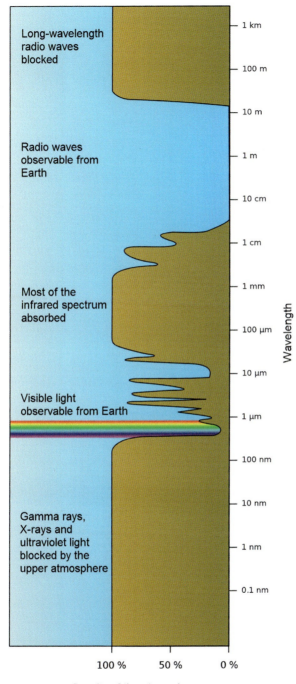

FIGURE 3.27 The opacity and transparency properties of Earth's atmosphere

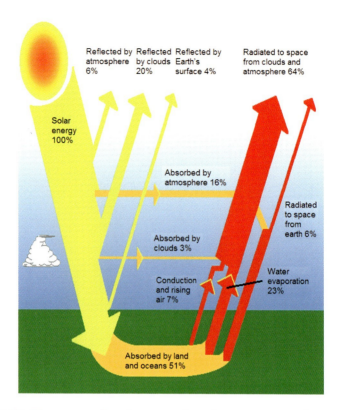

FIGURE 3.28 The atmosphere's properties of transparency determine the energy balance of Earth, the average temperature of which is currently about 15 °C. The balance presupposes equality between the incoming energy we receive from the Sun and that outgoing, radiated by Earth. If for any reason the input energy level is different from the output, the temperature varies to re-establish a new equilibrium. The energy radiated by Earth depends on its temperature. If it rises, the radiation increases

FIGURE 3.29 The color of the sea depends on the optical properties of the water

FIGURE 3.30 The Blue Marble, a famous photograph taken on December 7, 1972, by the crew of Apollo 17. The Sun was behind the astronauts and Earth, fully illuminated, stood out by virtue of its *blue color*

> **CONSIDER THIS**
> The Solar System is consists of eight planets, but only Earth is covered in liquid water. Why?

Space Telescopes, from Hubble Onwards

How can we roll back the boundaries of the visible universe? In the first place, with tools such as the telescope, the modern eyeglass.

Although the pupil, which collects the light in the eye, is very small, in a telescope the diameter of the objective that performs the same function is large, and is therefore able to collect much more light and to produce a magnified image of faraway objects.

The observation of celestial bodies is disturbed by the terrestrial atmosphere, which produces the characteristic flicker of light from the cosmos and degrades the image quality. This effect is created by air turbulence, and in order to reduce it, telescopes have often been placed in deserts or on mountains, where the air is fairly stable and dry. These locales make it possible to reduce flicker, but not to eliminate it. Is it possible to obtain an image without flickering if the effect of the atmosphere is completely removed. Absolutely. But how? Simple—by placing the telescope on a satellite.

The first and most famous space telescope is the Hubble Space Telescope (HST), put into orbit by NASA on April 24, 1990, and placed at a height of around 600 km (Figs. 3.31, 3.32 and 3.33). It owes its name to the great American astronomer Edwin Hubble, discoverer of the expansion of the universe.

Many discoveries have been obtained thanks to Hubble. We now have a more complete understanding of the expansion and age of the universe, of the presence and properties of blacks holes in the nuclei of galaxies, and of the characteristics of dark matter, which pervades the cosmos. With 342 exposures taken between December18 and 28, 1995, the telescope transmitted the Hubble Deep Field image to us, focused on such a small region as to include only a few stars in the Milky Way. The 3,000 light sources of the image are the most distant known galaxies. On February 21,2006, an astronomical object of unknown type was sighted, called SCP 06F6, perhaps a new type of supernova. Hubble has photographed the collision of a comet with the planet Jupiter and explored the transneptunian dwarf planets Pluto and Eris. It sent us the first image of an extrasolar planet, discovered several protoplanetary disks around forming stars in the nebula of Orion, and much more.

Hubble detects not only visible light but also infrared and ultraviolet radiation. Since the atmosphere absorbs most of the electromagnetic waves coming from the sky, only a space telescope can provide us with images of the universe outside the range of the visible.

Light and the Sky 129

FIGURE 3.31 Hubble's eyeglasses. Hubble had accidentally been designed to work in air, not in a vacuum; hence its images were distorted and out of focus. It was thus necessary to endow it with 'eyeglasses' capable of correcting this sight defect. In December 1993, a space mission of the space shuttle program with seven astronauts on board, reached it for a maintenance task and installed a system of corrective optics called COSTAR (Corrective Optics Space Telescope Axial Replacement), which allowed the telescope to operate at its best

This equipment was the first of a large family, now made up of dozens of instruments in orbit that continuously scan space, examining all types of radiation, from radio waves to gamma rays.

The cosmos as seen through non-visible radiation is full of surprises and an endless source of new discoveries. For example, even with the most powerful telescope, our eyes cannot see the center of the Milky Way due to the presence of cosmic dust that blocks visible light, while infrared radiation is able to cross it, allowing us to see it in detail.

130 Visible and Invisible

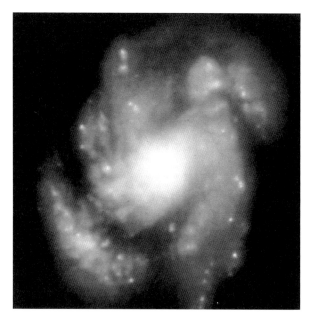

FIGURE 3.32 Images of the spiral galaxy Messier 100 provided by Hubble, before and after the installation of corrective optics

FIGURE 3.33 Images of the spiral galaxy Messier 100 provided by Hubble, before and after the installation of corrective optics

The Sun

The Sun, the mother star of our planetary system, continuously emits electromagnetic radiation and particles (solar wind) and is the foundation of life (Figs. 3.34, 3.35, 3.36 and 3.37). Without its presence Earth would be a lifeless pile of rocks. For a long time the source of the enormous energy generated by the Sun was not understood. Today, thanks to the theory of relativity, we know that it results from the transformation of the mass into energy according to Einstein's equation $E = mc^2$. In the nuclear reactions that occur in our star, 600 million tons of hydrogen are converted each second into 595.74 million tons of helium. Consequently, in the Sun 4.26 million tons of matter disappear each second, transformed into energy. In just one second, this mechanism produces the annual energy consumption of our planet multiplied by 800,000 (*Statistical Review of World Energy 2009*, 2008 data).

The Sun is a very complex system, not yet known in all its aspects. It is a star of small to medium size and has a radius of 696,000 km (about 109 times that of Earth) and a mass (about 333 times that of Earth) consisting predominantly of hydrogen (about 74 %) and helium (about 24 %). The percentages are those valid in our age. This process, which started 4.6 billion years ago, will continue for a similar length of time. Like all stars, the Sun is a living entity that is born, grows, and dies. Its stability arises from the equilibrium of two opposing forces—that of gravity, which tends to compress it, and a second force that tends to expand it, due to electromagnetic radiation continuously created by nuclear reactions.

The part of the Sun visible to us is its surface, called the photosphere, a region where the average temperature is about 5,500 °C. From this thin layer most of the Sun's solar energy is emitted in the form of ultraviolet, visible, and infrared radiation.

The solar corona lies around the photosphere, where the temperature is very high, around two million degrees. It is generally not visible, but it can be observed during a total solar eclipse, or through an instrument, the coronagraph, which artificially covers the solar disk. The corona produces a continuous emission of charged particles, primarily electrons and protons. As a result of this flow, called the solar wind, which spreads far beyond the orbit of Earth, the Sun loses about 800 kg of matter every second.

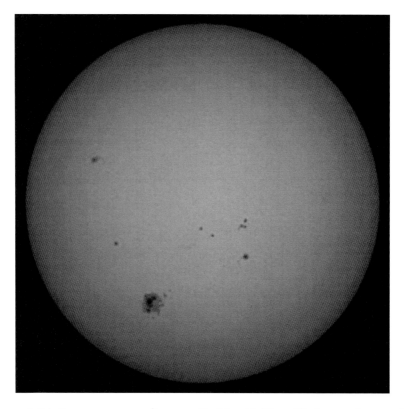

FIGURE 3.34 Sunspots were first studied by Galileo. Today we know that these dark areas on the photosphere are cooler than the surrounding areas and are associated with strong magnetic fields

Between the photosphere and the corona lies the layer of the chromosphere, analyzed recently by the solar optical telescope aboard the Hinode probe, a collaboration between Japan, the United States, the United Kingdom, and Europe (Fig. 3.35).

In the innermost part, the core, pressure and temperature can reach very high values, almost 500 billion atmospheres in pressure and about 15 million degrees in temperature. Under these extreme conditions, nuclear fusion reactions occur spontaneously, producing energy.

This very intense activity of the Sun produces a strong magnetic field and phenomena such as sunspots (Fig. 3.34), flares (bright spots), and the prominences (huge luminous jets).

CONSIDER THIS
Why is sunlight white?

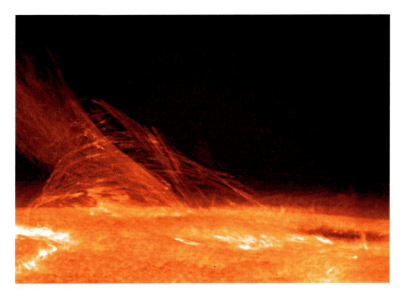

FIGURE 3.35 Details of the intense dynamics of the chromosphere transmitted by the solar optical telescope aboard the Hinode probe on January 12, 2007

FIGURE 3.36 X-ray emission by the Sun. Since our atmosphere absorbs this radiation, we need to use a satellite to detect it. These rays are invisible, and to represent them, the 'false-color' technique is adopted: the wavelengths were reduced, shifted into the range of visible light, consequently different colors correspond to different wavelengths of X-rays

FIGURE 3.37 The Sun's spectrum. Since Newton's day, we have known that a prism can break down sunlight into different chromatic components, but only in the early nineteenth century it was realized that the solar spectrum is crossed by several *black lines*. Since each *color* corresponds to a specific wavelength, why do these 'holes' appear? In other words, why do we not receive the light with a wavelength corresponding to the *black lines*? Today we know that these *lines* are due to atoms absorbing the 'missing' colors, and, by studying this effect, we can determine which gases were traversed by the light of the Sun, particularly those of the solar atmosphere

KISSED BY THE SUN
The sound of a Stradivarius violin is inimitable, without equal. Why?

According to some, the answer may lie in the activity of the Sun. Stradivari was born in 1644, a period well known to experts of our star. In the years between 1645 and 1715 sunspots, which usually appear on a regular basis, suddenly disappeared. This period was called Maunder minimum and coincided with the central and colder part of the so-called Little Ice Age, during which the winters were extremely cold. This suggests a correlation between the activity of the Sun and the changing climate.

The long winters and cold summers of the time created unique environmental conditions for the Norway spruces of the Alps, used by luthiers for the construction of violins, providing them with wood that provided an incomparable sound.

Eclipses

When the Moon stands between the Sun and Earth, so as to obscure the solar disk, there is a solar eclipse. On the contrary, if our own planet projects its shadow onto the Moon, we have a lunar eclipse.

The total solar eclipse occurs when the Moon completely covers the solar disk and the three celestial bodies are perfectly aligned.

Light and the Sky 135

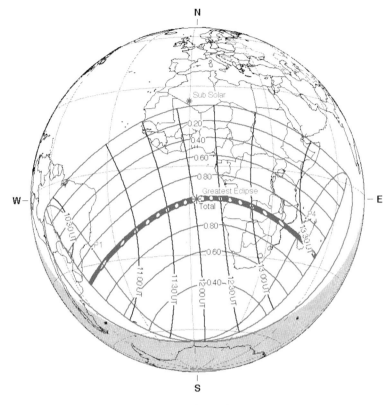

FIGURE 3.38 The path of the total solar eclipse, June 21, 2001

This phenomenon is rare. Any given point on the surface of Earth sees a total eclipse on average every 370 years. The perfect alignment of the three bodies is visible only along a narrow path, while in adjacent areas a partial eclipse may be observed (Figs. 3.38 and 3.39).

Since ancient times, humans have measured the motions of the Sun, Moon, and stars. The ability to predict the regular and periodic motion of celestial bodies conferred security and power to the ancient civilizations, but this all went awry in the face of an unexpected episode, such as the eclipse. Seeing the Sun disappear in full daylight was frightening and might foretell catastrophic events (Fig. 3.40).

The most advanced civilizations succeeded in predicting eclipses, thus removing their status as random events. The Chaldeans, about 2,500 years ago, realized that after 223 lunar months, the Sun, Moon, and Earth find themselves in the same relative positions; therefore, any eclipse would occur again after 223 lunar months Thus the Saros cycle came about, during which 29 lunar eclipses and 41 solar eclipses take place.

136 Visible and Invisible

FIGURE 3.39 A photo of the total eclipse of June 21, 2001, taken in Lusaka, Zambia. The corona, normally obscured by the brightness of the Sun, appears clearly visible during the eclipse

> **OBSERVING ECLIPSES SAFELY**
> Under normal conditions, sunlight is so intense that it is difficult to look at it directly, which is fortunate, because observing the Sun, even for a few seconds, may result in damage to the retina, an organ insensible to pain. Looking at the Sun during an eclipse is equally dangerous, except during the short period of total coverage, in the case of total eclipse. NASA has simple instructions on its site on this matter: "Even when 99 % of the Sun's surface is obscured during the partial phases of a total eclipse, the remaining photospheric crescent is intensely bright and cannot be viewed safely without eye protection. [...] The Sun can be viewed directly only when using filters specifically designed for this purpose."

Light and the Sky   137

FIGURE 3.40 One of the oldest representations of a total solar eclipse comes from the Neolithic tombs of Loughcrew (Ireland) and dates to the early agricultural settlements of that land, around 3300 B.C

Stars and Galaxies

The light that reaches Earth does not come from the Sun alone but also from the other stars that crowd the sky. The fact that the solar light is dominant compared to that of the stars, and that these may be seen clearly only at night, is a well-known paradox.

The colors of the stars in fact vary. For example, the white of Sirius is different than the red of Betelgeuse. It was understood early on that the color of a star depends on the temperature of its surface. Blue indicates high temperatures, yellow denotes an intermediate temperature, and red stars have the lowest temperature. Subsequently, it was realized that the surface temperature of stars is related to their absolute brightness, obtained after eliminating the effect of the distance between source and observer.

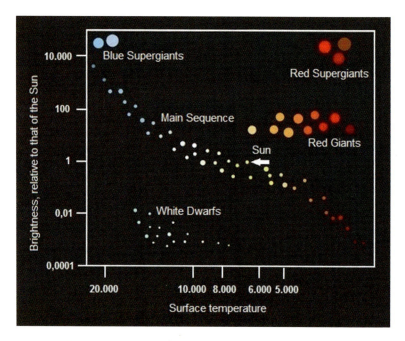

FIGURE 3.41 The Hertzsprung-Russell diagram correlates the surface temperature of stars to their brightness

The ensuing graph, called the Hertzsprung-Russell diagram, named after the two astronomers who proposed it, is helpful to understand the evolution and characteristics of the stars (Fig. 3.41).

Most stars are located in a well-defined region, called the main sequence, where they stay for most of their existence, and at this stage they produce energy by converting hydrogen into helium. This band extends from the blue stars, the brightest and hottest, to the red dwarfs, weak and cold. The Sun is a yellow dwarf and lies in the central part of the sequence.

Towards the end of their lives, stars leave the main sequence, with different fates depending on their mass. On the following pages we will explore some of these final stages.

The red supergiants, with a mass exceeding 10 solar masses after transforming all the hydrogen into helium, begin to fuse the latter into heavier elements. These stars display rather low surface temperatures, yet they have enormous dimensions. Many have radii 200 to 800 times greater than those of our Sun. If they were located where the Sun is in our Solar System, they would go

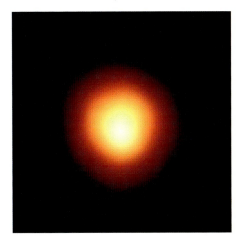

FIGURE 3.42 The first direct image of the disc of a star other than the Sun, Betelgeuse, photographed using the Hubble Space Telescope in December 1996

beyond Earth's orbit. A well-known red supergiant is Betelgeuse, the second brightest star in the constellation of Orion (Fig. 3.42). Astronomers believe that it will conclude its existence by exploding in a brilliant supernova.

The blue supergiants, despite having a large mass (like the red ones), have a very high surface temperature. These rare stars are the brightest of all in the universe and are believed to have originated from the evolution of red supergiants in which nuclear fusion has decreased, causing a contraction with a consequent increase of surface temperature. A famous blue supergiant is Rigel, the brightest star in the constellation of Orion.

Stars of a smaller mass (less than 10 solar masses), after finishing the fusion of hydrogen, become red giants. These stars also have rather low surface temperatures. An example is Aldebaran, the brightest star in the Taurus constellation.

Red giants, after finishing the fusion of other elements, become white dwarfs, celestial bodies without an autonomous source of energy, with a temperature that gradually decreases; they exhibit small size and high density (and therefore gravity). The Sun will become a red giant in about 5 billion years, and eventually a white dwarf.

FIGURE 3.43 The Milky Way is a barred spiral galaxy, i.e., made up of a nucleus with two extensions of stars, which as a whole are reminiscent of a bar, from which the arms of the spiral branch off. Obviously, it is not possible to observe it from the outside. To get an idea of what it looks like, we may examine clusters of a similar shape, such as the NGC 1300, located 69 million light-years away

Nearly all the stars are grouped in large clusters. The Milky Way, the whitish bright band with a milky appearance that diagonally crosses the celestial sphere, is simply the galaxy to which the Sun belongs, and it contains about 300 billion stars (Fig. 3.43).

The Milky Way is not alone, as was thought for a long time. Thanks to the Hubble Space Telescope, in particular, it has been possible to estimate that in the observable universe, there are about 125 billion galaxies. These are not scattered at random; many are grouped into clusters containing up to 1,000 of them.

According to the Big Bang model, the cosmos on a large scale is the same regardless of the position from which it is viewed. This hypothesis, known as cosmological principle, foresees a homogeneous universe (independent from the point of observation) and isotropic (independent from the direction taken into consideration), similar to a dilute gas, in which galaxies take the place of molecules.

Why Is the Sky Dark at Night?

In 1692 the Reverend Richard Bentley (1662–1742) asked Newton a question that may be reformulated in these terms: If stars and planets attract one another as a result of gravitational force, why they do not precipitate towards some common center, condensing into a large mass?

Gravitational collapse, Newton replied, does not happen in an infinite universe. A center can only be found in a finite volume, so if matter is distributed in an unlimited space, there is no point on which to converge. Not only that, but in Newton's theory, the stars are uniformly distributed; that is, their density is constant. Each body is attracted in every direction by equal forces that are balanced, generating a condition of equilibrium. According to the great physicist, therefore, our universe is infinite, homogeneous, and static (Fig. 3.44).

FIGURE 3.44 The Sagittarius star cloud, a vast cluster of stars visible in the constellation of Sagittarius, in a photograph taken by the Hubble Space Telescope

The Newtonian picture of the universe, unlimited and not subject to evolution, posed several problems. In particular, it did not explain why the night sky is dark.

This incongruity was formulated as a paradox by the physicist Wilhelm Olbers (1758–1840) in the first half of the nineteenth century. If stars shine with a constant light and are uniformly distributed, the whole sky in a static universe, infinitely old and infinitely extended, must appear uniformly bright. The paradox was based on the following observations:

- Light is spread throughout the whole of space, decreasing in intensity when its source moves away from the observer. In particular, if we consider two identical stars but place one at a distance double that of the other, the light that comes from the further away star is a quarter of that received from the nearby star. Luminous intensity varies proportionally to the inverse square of the distance.
- The number of stars around Earth increases with relative distance, in such a way as to counterbalance the decrease in intensity. In fact, the number of stars present in a spherical layer varies in proportion to the square of the distance from the center (Fig. 3.45).

The key point of the Olbers' argument was that the amount of light that comes from a layer of stars is always the same, whatever the distance that separates them from us, since with increasing distance, the decrease of light intensity and the increase in the number of stars are balanced. Consequently, since the universe around Earth may be imagined as consisting of an infinite number of concentric layers, the light on our planet would be huge, illuminating both the day and night.

As far back as 1744. The Swiss mathematician Jean-Philippe Loys de Cheseaux (1718–1751) developed a rigorous treatment, considering a finite although incredibly large space. This analysis showed that in a spherical universe, with a radius equal to 3,000 billion light-years, the brightness of the stars would be 180,000 times more intense than that of the Sun.

The paradox may be overcome by abandoning the idea that the universe is static, homogeneous, and infinitely old.

The first to find a way to eliminate this incongruence was not a scientist but a writer, Edgar Allan Poe (1809–1849), in his

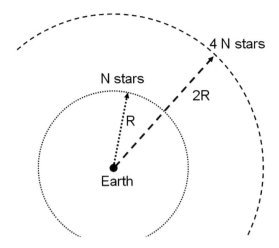

FIGURE 3.45 In the mid-nineteenth century, it was thought that the universe was static, homogeneous, and infinite, consisting of concentric layers with Earth at the center. Consequently, the number of stars contained in each spherical layer had to vary proportionally to the square of their distance from Earth

prose-poem "Eureka" in 1848: "Were the succession of stars endless, then the background of the sky would present us an uniform luminosity, like that displayed by the Galaxy -since there would be no point, in all that background, at which would not exist a star. The only mode, therefore, in which, under such a state of affairs, we could comprehend the voids which our telescopes find in innumerable directions, would be by supposing the distance of the invisible background so immense that no ray from it has yet been able to reach us at all." (Fig. 3.46).

This argument is relevant for two reasons:

- It takes into account the finite speed of propagation of light.
- It supposes an evolving universe, in which the luminous rays have not yet reached us.

Poe's reasoning is essentially correct.

How about Olbers' paradox today? The darkness of night, according to current opinion, stems from three factors:

- The speed of light is finite.
- The universe is not infinitely old.
- The universe is expanding.

144 Visible and Invisible

FIGURE 3.46 Edgar Allan Poe in a painting by Oscar Halling in 1860, using an 1849 daguerreotype as a model

The first two points imply that Earth receives light from a finite number of stars, those included in a sphere of radius equal to the distance traveled by light from the birth of the universe until today. The third, on the other hand, implies an increase in the wavelength of the luminous rays received from distant stars. This modifies their properties, with a shift of radiation from the visible interval to that of the invisible. None of the three effects, taken individually, may resolve the paradox, which may only be explained when considered all together.

Light-Years

Light travels very fast, covering 300,000 km in 1 s (in a vacuum). But then how far does it travel in a year?

A year consists of 365 days, a day 24 h, and one hour 3,600 s; therefore a year is equivalent to $3,600 \times 24 \times 365 = 31,536,000$ s. Consequently, in a year light travels:

$$31,536,000 \text{ s} \times 300,000 \text{ km/s} = 9,460,800,000,000 \text{ km}$$

by considering more precise values for both the speed of light and the duration of the year we obtain the value 9,460,730,472,580.8 km.

It covers an enormous distance, which is used as a unit for measuring cosmic space, the light-year, which indicates the distance traveled by light in 1 year and corresponds to approximately 9,461 billion km (Table 3.1).

Moving away from our galaxy, distances increase considerably and are measured in thousands, millions, and billions of light-years (Fig. 3.47).

The first galaxies that we encounter are the Magellanic Clouds, two small groups that orbit around the Milky Way as satellites. The Great Magellanic Cloud is 157,000 light-years away, while the Small Magellanic Cloud 200,000 light-years away.

At even longer distances, the great Andromeda Galaxy is located 2.5 million light-years from us. The most distant object visible to the naked eye, under optimal conditions, is the Triangulum Galaxy, 3 million light-years from Earth.

The Milky Way, Andromeda, the Triangulum Galaxy, as well as a series of other smaller galaxies, are located in the so-called Local Group, which extends across a region 10 million light-years wide.

The Local Group in turn is part of a much larger structure, the Virgo Supercluster or Local Supercluster, shaped like a flattened disk with a diameter of 200 million light-years. It includes about one hundred groups and clusters of galaxies, and its center lies 60 million light-years away from us.

One of the most distant galaxies, identified in 2006 and denoted by the symbol IOK-1, is 12.88 billion light-years away. The images we receive from it are those emitted by a celestial body 12.88 billion years ago. The most accurate estimate of the

TABLE 3.1 Some distances relating to the solar system and the Milky Way

Distance	Time spent by light	Distance in light years
Earth to Moon	1.25 s	40×10^{-9}
Sun to Earth	8 min 19 s	16×10^{-6}
Sun to Saturn	1 h 19 min 20 s	0.5×10^{-3}
Earth to Proxima Centauri	4.22 years	4.22
Sun to the center of the Milky Way	26,000 years	2.6×10^4
Radius of the Milky Way	About 100,000 years	About 10^5

Proxima Centauri is the star closest to the Sun

age of the universe is 13.82 billion years, so the image we received was probably that of a galaxy shortly after the Big Bang, when the cosmos was just 940 million years old.

> **ASTRONOMICAL UNITS OF MEASUREMENT**
> The light-year is not the only unit of measurement of cosmic distances. The distance between Earth and the Sun is called the astronomical unit (au), corresponding to 1.496×10^{11} m = 149.6 million km and thus less than the light-year, the length of which is approximately 9,461 billion km and equivalent to approximately 63,240 au. A second unit of measure, greater than the light-year, is the parsec, equal to 30,857 billion km and so equivalent to about 3.26 light-years, or 206,263 au.

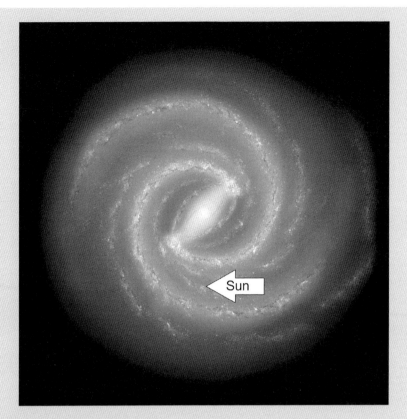

FIGURE 3.47 An artistic representation of the position of the Sun in the Milky Way. Its location in a habitable galactic zone is evident, far from the center and yet also far from major sources of radiation

Space Travel

There are many questions that arise in the face of the immensity of the cosmos. Are there other planets like Earth? Are there other living beings in the universe? If so, are they intelligent beings?

So far, the only celestial body visited by people is the Moon, where between 1969 and 1972 twelve astronauts of six different Apollo missions landed. The Moon, like the other bodies (planets, satellites, and asteroids) of the Solar System is inhospitable to life, but despite this, the development of space travel will take place by

148 Visible and Invisible

FIGURE 3.48 This is a representation of a hypothetical space colony that the Californian artist Donald Davis produced for NASA

exploring these territories. There are already several projects for settlements based on the creation of closed ecosystems in which some kind of human activity is possible (Fig. 3.48).

However, the real challenge for humanity is not merely traveling within the Solar System. Our galaxy, the Milky Way, is populated by at least 300 billion stars. Why stop at the Sun?

Many celestial bodies may be similar to Earth and therefore difficult to detect because they do not emit light. Some of these, called extrasolar planets, or exoplanets, have been identified, and the search for extraterrestrial life forms has begun (Fig. 3.49).

Various estimates have been formulated as to the number of civilizations hypothetically present today in the Milky Way. Of course these assessments are hardly definitive, but a plausible result speaks of 600 possible civilizations. This calculation lies at the basis of the numerous ongoing attempts to communicate with extraterrestrial intelligences through the use of electromagnetic waves.

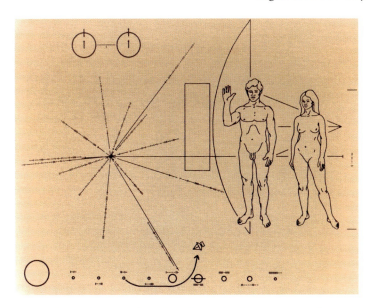

FIGURE 3.49 The plaque placed onboard the Pioneer 10 spacecraft, launched on March 3, 1972, containing a variety of information on our civilization, in the event that it was intercepted by extraterrestrial beings. In particular, the drawing of a man and a woman, whose dimensions are obtainable from a comparison with the representation of the spacecraft, the position of Earth in the Solar System, and that of the Sun relative to the center of the galaxy and to 14 pulsars

Since radio signals travel at the speed of light, sending them into space is a quite different matter from sending spaceships. The crucial point is that space in the galaxy and in the whole universe is almost entirely empty: on average, stars and galaxies are very far apart, therefore among the many difficulties related to space travel, the main one concerns the time required to accomplish them.

The traveling time required to go to the Moon is at least 4 days. This seems little, but light takes only 1.25 s to reach our satellite.

The nearest star to the Sun is Proxima Centauri, 4.2 light-years away. If light takes all this time to reach it, how much time might a spaceship take? Suppose you travel, following a straight trajectory and using a chemical-fuel vehicle, at 40,000 km/h, a speed not reached thus far. In spite of these optimistic assumptions it would take us a good 113,400 years to reach our destination.

For this reason, with current scientific and technological knowledge, space travel beyond the solar system appears impossible. Will it always be so? Some scientists are convinced that the future will provide different perspectives.

The British scientist Stephen Hawking, in an interview on December 1, 2006, stated in the online version of the Daily Mail that "the long-term survival of the human race is at risk as long as it is confined to a single planet. Sooner or later, disasters such as an asteroid collision or nuclear war could wipe us all out. But once we spread out into space and establish independent colonies, our future should be safe. There isn't anywhere like the Earth in the solar system, so we would have to go to another star."

Suppose you are traveling just under the speed of light, let's say at 200,000 km/s. In this case, Alpha Centauri could be reached with a spaceship in 6 years. This would be the duration of the voyage for those who remain on Earth, while for the crew the time would be reduced to 4.5 years due to relativistic effects.

Is it possible to travel at such a high speed? Nothing prevents it, except for the huge amount of energy needed. Hawking proposes considering antimatter as fuel. If, in the engines of the spaceship, antimatter could be made to interact with matter, it could generate the energy required.

In the distant future, perhaps all this will be possible. If this happens, its precursor will not have been a scientist, but a writer. What Hawking proposes is in fact the technology used by the engines of the starship Enterprise, as featured in the science fiction series *Star Trek*, conceived by Gene Roddenberry (1921–1991).

Antimatter

First theoretically predicted, and then discovered through experimentation, antimatter is now a certainty. For each particle with mass, like the electron or the proton, there exists an equivalent antiparticle. When antimatter meets matter, the result is dramatic. Both vanish (are annihilated), creating light, on the basis of Einstein's equation $E = mc^2$.

The same transformation of matter into energy takes place in nuclear reactions, but there is a significant difference. In the latter only a small part of matter is converted into energy, while in annihilation, all matter and antimatter is transformed.

As far as we know today, there is no antimatter in the natural state. It was annihilated by the encounter with matter in the first instants of the Big Bang, a phenomenon that flooded the cosmos with that light that came down to us in the form of the cosmic background microwaves. For some reason still unknown to us, the amount of matter was greater than that of antimatter, although very slightly—one part in a billion more. As a result, all the antimatter and almost all matter annihilated one another. If there had not been this slight difference, the universe would now be full of light without matter. It's thanks to this excess part that we now exist.

Antimatter is created in research laboratories, particularly in particle accelerators. There is already a technological application for it. The Positron Emission Tomography (PET) is a tool for medical diagnostics that utilizes the emission of anti-electrons (positrons) to achieve high-resolution images of the internal organs of the human body.

Supernovae

Every star is born, lives, and dies. The duration of the sequence depends on the stellar mass. The greater the celestial body, the shorter the period of its existence. The evolution of a star initially follows a common path, differentiating itself in the final phase, when its hydrogen runs out.

Low-mass stars will begin to become extinct by expanding to become red giants, and then finish their existence as small celestial bodies, the ever-cooling white dwarfs. The perspective for the massive stars, the brightest, is very different. The cycle of nuclear fusion allows them to produce the nuclei of many atoms internally: helium, carbon, nitrogen, oxygen, neon, magnesium, silicon, calcium, and iron. The mechanism is simple. When the process of fusing hydrogen into helium finishes, it starts to convert the latter into carbon and so on. Once it reaches iron the process stops, because the fusion of the iron nucleus does not produce energy but requires it.

FIGURE 3.50 The light emitted by a supernova may be compared with that diffused by an entire galaxy. The Hubble Space Telescope sent us this image of the supernova SN 1994D (*bottom left*) and its galaxy, the NGC 4526

When the mechanisms of nuclear fusion are no longer able to produce energy, an enormous explosion takes place, called a supernova, with the dispersion into space of the outer layers of the star. This event produces an amount of energy in a few days comparable to that radiated by our Sun over billions of years, and a light emission comparable to that of the entire galaxy that hosts it (Fig. 3.50).

Thanks to the enormous energy available during the explosion, nuclei heavier than iron are formed, including cobalt, uranium, nickel, lead, iodine, tungsten, gold, and silver.

The supernova disperses into space both light atoms, created by the star, and those synthesized by the explosion. This is its greatest gift. We terrestrials exist thanks to starbursts that occurred billions of years ago, before the origin of the Sun and the planets. Our Solar System was formed by the condensation of an interstellar

Light and the Sky 153

FIGURE 3.51 The remnants of Supernova N63A in the Large Magellanic Cloud, as photographed by the Hubble Space Telescope

cloud containing the atoms of elements that we know, which originated from previous supernovae (Fig. 3.51).

When such an explosion occurs in our galaxy, it is visible to the naked eye, meaning that the celestial sphere is enriched by a very bright (*super*) new star (*nova*), hence the term 'supernova.'

In antiquity, the extremely rare appearance of a new star, which later faded away, constituted a truly exceptional phenomenon. The last supernova observed in the Milky Way was the one that appeared in 1604, known as Kepler's Supernova (or Kepler's Star).

Black Holes

In the explosion of a supernova, the outer part of the star is dispersed into space. Let us now consider the evolution of the residual inner part. The force of gravity shrinks it until it deeply changes its nature.

In atoms, almost all the mass is concentrated in the central part, in the tiny nucleus that contains protons and neutrons. The size of the atom is dictated by the cloud of electrons surrounding the nucleus, which occupies a region far larger than the nucleus itself. If we magnified an atom one trillion times, it would take up the space of a football field (about 100 m), and its nucleus would be the size of a pinhead (1 mm). The atom is thus essentially empty; the low densities of familiar substances, from water to iron, are due to this feature.

In the core of a supernova, the compression melts protons with electrons to form neutrons, which combine with those already present. The result is a stellar core of only neutrons, with the empty space characteristic of atoms no longer around it. It is as though these atoms, as big as football fields, were compressed until they became no larger than pinheads.

Under these conditions, the concentration becomes extremely high. A small block of 1 cubic centimeter would acquire a mass of a billion tons. Thus, a neutron star is born.

On its surface, the force of gravity is enormous—up to 100 billion times more intense than that found on Earth. So it is very difficult to get away from a neutron star!

If the core remaining after the explosion is large enough, at least three times the mass of the Sun, the gravity is so strong that nothing can escape it, not even light. In this case a black hole is formed (Fig. 3.52).

Neutron stars are more numerous than black holes; it is believed that our galaxy contains a billion neutron stars and several million blacks holes.

How can we spot a black hole, if it does not emit light? By its strong gravity. Sometimes one of these objects absorbs a nearby star, and we see the light emitted by matter falling down into it. Gravity also determines the speed at which gas clouds rotate around the body. Some blacks holes have recently been identified because they are coupled with a star, thus forming a binary system. Only the star is visible, but its rotation and the characteristics of the radiation emitted indicate the presence of a black hole interacting with it.

Light and the Sky 155

FIGURE 3.52 This is an artistic representation of a black hole, as envisioned by NASA, in which the curvature of the light rays coming out of the core, then sucked back by gravity, is represented. The deviation of the light due to gravity was foreseen by Einstein's relativity theory

In 2007, the Chandra X-ray space observatory detected a black hole with a mass 15.7 times that of the Sun belonging to the galaxy of the Triangle and about 3 million light-years from Earth. The black hole, called M33 X-7, orbits around a companion star that eclipses it every 3.5 days, thus forming a binary system (Figs. 3.53 and 3.54).

TIME NEAR A BLACK HOLE
A black hole absorbs everything, but at a certain distance its force of gravity varies like that of a normal star. When, approaching a black hole, you exceed a specified threshold, called the event horizon, and you are inexorably dragged towards its interior.

FIGURE 3.53 An artistic representation of black hole M33 X-7, which rotates together with its companion star (the *blue* object), about 70 times larger than the Sun. The *black hole* lies at the center of the *orange* disk: a structure that represents the matter in orbit around the *black* hole prior to being absorbed

If, onboard a spaceship, you could remain on the sidelines of the event horizon, the extremely strong gravity would distort the flow of time: in the presence of a black hole with a mass equal to one thousand times that of the Sun, time would flow 10,000 times slower, and thus a year spent in such conditions would be equivalent to 10,000 years on Earth.

The Active Galactic Nuclei

The brightest bodies in the cosmos are not supernovae but the active galactic nuclei, the most powerful source of electromagnetic radiation in the universe. Such extremely strong emission can take place over the entire electromagnetic spectrum—radio waves, infrared, visible light, ultraviolet, X-rays, and gamma rays.

The first discovery of this kind took place in the field of radio waves. In the 1960s, objects were identified, billions of light-years

FIGURE 3.54 Graphic that represents data provided by two different satellites: the orbital telescope Chandra, for the observation of X-rays, and the Hubble Space Telescope. The *blue* light in the center represents the X-rays emitted by the binary system made up of the black hole M33 X-7 and its companion star, while the *bright spots* that surround it are young stars. The variation of X-rays due to rotation allows us to estimate the size of the *black hole*

away, that emit radio waves in large quantities, from a hundred to a thousand times greater than that released by an entire galaxy. The first sighting dates back to when it was realized that what appeared to be a signal from a faint star, indicated by the abbreviation 3C 273, actually came from a very distant object, 2.4 billion light-years away. No star could be so powerful as to convey its light over such a distance. These bodies were given the name quasars, an acronym for "quasi-stellar radio source." For a long time their nature was questioned. It is now thought that they belong to the class of active galactic nuclei, which not only includes quasars but also other extremely intense light sources.

The amount of energy irradiated by active galactic nuclei is enormous, equal to that generated by trillions of stars like the Sun. It is believed that this emission is caused by matter falling into a

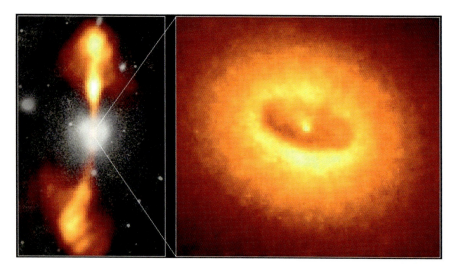

FIGURE 3.55 The photo on *left* shows a composition of optical images (in *white*) and radio (in *orange*). The disk of galaxy NGC 4261, consisting of hundreds of billions of stars, appears *white*, while the radio image, which shows a couple of jets about 88,000 light-years long is displayed in *orange*. In the photo to the *right*, the optical image of the central part of the galaxy is shown, the active galactic nucleus that rotates around the supermassive *black hole*

supermassive black hole, one whose mass exceeds that of the Sun by one million to one billion times (Figs. 3.55 and 3.56).

Let us now consider two systems, known as being active galactic nuclei. NGC 4261, analyzed by the Hubble Space Telescope, is one of the twelve brightest galaxies in the Virgo cluster, 45 million light-years away. Its central part has the shape of a disk and constitutes the active galactic nucleus that feeds the black hole, where gravity compresses and heats the material. The extremely hot gases, diffused nearby, produce two radio jets detected by radio telescopes through their characteristic double-lobe structure, which extends far beyond the galaxy.

Let us now consider the second case, also studied thanks to Hubble. In 1997 the speed of rotation of an enormous gas disk was measured, with a radius of 26 light-years, rotating around the center of galaxy M84. The latter is visible in the Virgo constellation and is 60 million light-years from Earth. From the speed of rotation, it has been deduced that the mass of the object around which

FIGURE 3.56 Image provided by the Hubble Space Telescope of the barred spiral galaxy NGC 4319 and the quasar Markarian 205. NGC 4319 is located at a distance of about 80 million light-years from Earth, while MRK 205 is much farther away, about a billion light-years

the gas cloud turns is equal to 300 million solar masses. Only a black hole can contain so much matter in such a small space. This is indirect evidence but convincing.

What happens when a supermassive black hole has incorporated all the mass present in its vicinity? Light emission ceases and the galaxy goes from active to normal. It is thought that in every galaxy a supermassive black hole is present, growing in active ones, at rest in normal ones, such as in the case of the Milky Way.

380,000 Years Since the Big Bang

When we see the light of a lamp or of a star, the light beam, which travels from the source up to our eyes, meets up with atoms in its path. These atoms may interact with the light by absorbing it all or in part. In the first case, the beam is suppressed, in the second, it is weakened. Since the universe is dominated by empty space, a vacuum, lots of light can travel undisturbed for millions or billions of years, carrying the images of stars and galaxies (Fig. 3.57).

Scientists do not think it has always been that way. According to the Big Bang model, the only one able to incorporate all the existing knowledge in a coherent picture, time and space began about 14 billion years ago. The whole universe was concentrated in an infinitely dense and hot point—the initial singularity. The process of expansion, still ongoing, had its origin then. Outside the regions occupied by the cosmos, there is nothing. Matter, light, energy, space, and time are inextricably linked to our universe, which is constantly evolving.

Like a normal gas, even the cosmos cools by expanding. The phase of expansion and cooling continues, even today, by steadily increasing the amount of space. This is the same phenomenon that magnifies the writing on a balloon when it is inflated. This is why we see all the galaxies moving away. They are not distancing themselves from Earth; rather, it is space that continues to expand.

We do not know what happened before the Big Bang. But we can describe what happened between the first moments until today.

A large number of photons, in the form of high-energy gamma radiation, was present in the very first seconds. After 2 min from the Big Bang, the temperature dropped a few hundred million degrees and the first nuclei of hydrogen and helium were formed—the central part of the future atoms. The temperature was still too high, though, and the electrons could not bind to the nuclei, being constantly bumped into by photons. The same light that blocked the formation of atoms was unable to travel, since electrons filled the cosmos. The continuous deviation of light rays by the latter would have made it impossible to view any object. You might have perceived only a diffused brightness, a very bright and shiny fog.

Finally, 380,000 years after the Big Bang, the temperature of the universe descended to below 3,000 K. Under these conditions,

FIGURE 3.57 An infrared view of the Milky Way taken by the COBE satellite during its mission to test the Big Bang theory. Our galaxy emits infrared light, invisible to the naked eye but detectable through appropriate instruments

the energy of the photons was not enough to tear electrons apart from nuclei. Neutral atoms were formed, less reactive than electrons. At this point, light was able to freely travel in space, separating itself from matter and giving rise to cosmic background radiation.

That radiation was spread over several wavelengths, not randomly distributed but determined by collisions with electrons. The overall effect produced a black body spectrum at a temperature of 3,000 K.

This has remained essentially unchanged since then, except for one aspect – its wavelengths have increased a thousand fold. This effect is due to the expansion of the universe, which has expanded in every direction by a factor of 1,000 (Fig. 3.58).

The Kelvin (K) unit of temperature measurement is obtained by adding the constant 273 to degrees Celsius (°C); 273.15 for

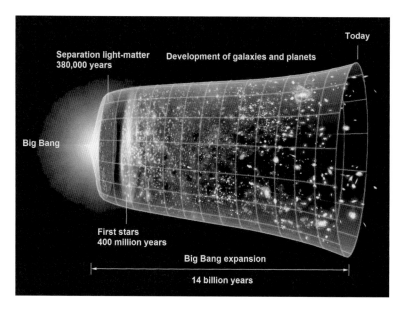

FIGURE 3.58 Schematic evolution of the universe, from the Big Bang to the present day

precise calculations. For example, 27 °C corresponds to about 300 K. If the temperature is expressed in degrees Fahrenheit, °F, two operations are required:

1. add the constant 459.67 and
2. multiply the result you get by 5/9 (an interval of 1 °F corresponds to 5/9 of a Kelvin).

Although the original radiation was in the region of visible light and infrared, today the emission lies in the microwave field, with a maximum peak of around $\lambda = 1$ mm. In terms of temperature, the initial 3,000 K was reduced by a factor of 1,000, reaching a value of about 3 K, just 3° above absolute zero.

Penzias and Wilson measured this radiation precisely, which carries traces of the properties of the universe at the time of formation of the first atoms, 380,000 years after the Big Bang. Their pioneering data were confirmed by subsequent measurements made with various instruments, including satellites and weather balloons, which established the astounding correlation of the measurements and the theoretical spectrum of a black body with a temperature of 2,725 K.

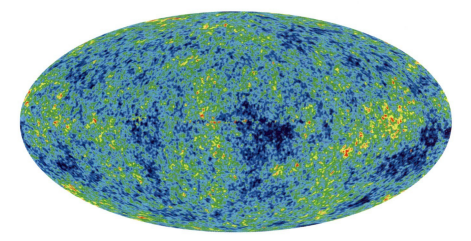

FIGURE 3.59 The anisotropy of background radiation. Recently it has been possible to measure the very small anisotropies (changes due to monitoring direction) of the cosmic microwave background radiation. The figure shows the distribution of radiation across the sky. The data are projected onto a plane, using the techniques developed to describe Earth's surface. On the map, the fluctuations are indicated by different colors, capable of showing variations of 0.0002°. The average temperature is 2.725 K; the *red* regions are warmer, the *blue* ones cooler. These results provide us with information on the early universe, and in particular testify that the distribution of atoms was not completely homogeneous. It is believed that the ripples in the density of matter resulted to the structures that populate the cosmos today—clusters of galaxies and vast empty regions

The existence of the cosmic microwave background radiation provided decisive confirmation of the Big Bang theory, and not only that. Radiation is considerably isotropic, substantially independent of the direction in which you point the instrument. The isotropy is very high. The difference between signals taken in two distinct directions is the order of ten millionths of a degree. Furthermore, this property is the confirmation of a theoretical hypothesis, linked to the Big Bang, known as the cosmological principle.

The first significant results on the anisotropy of radiation were obtained thanks to the scientific COBE satellite, launched by NASA in 1989, and studies of the data carried out by two of the main designers, John Mather and George Smoot, for which they were awarded the Nobel Prize. The study of cosmic microwave background radiation has continued through increasingly precise detections using balloons as 'boomerangs' or satellites such as the American WMAP and the European PLANCK (Fig. 3.59).

4. Light and Life

> *"The happiness of the bee and the dolphin is to exist. For man it is to know that and to wonder at it."*
>
> Jacques-Yves Cousteau, researcher and oceanographer

The Evolution of the Eye

The ability to make out an image using the visual system has developed in almost all species of the animal world. It is believed that all eyes have a common origin—a proto-eye, formed about 540 million years ago.

Some animals, such as spiders and insects, still use primitive eyes to detect light. On their head, close to their two main eyes, are located two or three light receptors, called ocelli (Fig. 4.1).

The ocelli do not detect images, but simply the intensity and polarization of light. This latter property is very important for orientation, since it allows the species equipped with ocelli to locate the Sun's position even on a cloudy sky.

The human eye also developed from luminous receptors. The first eyes were mere light receptors, similar to those of taste and smell. They could detect the intensity but not the direction of a light ray (Fig. 4.2). The receptors gradually sank into a cavity, thus becoming able to partially discriminate the direction of light (Fig. 4.3).

The increase in the size of the cavity and the reduction of its opening generated what was in many ways a camera obscura (Fig. 4.4). Then, the development of a layer of transparent cells on the opening created a closed cavity, which filled with liquid, known as vitreous humor (Fig. 4.5). Thereafter, the protective cells divided to form the crystalline lens (Fig. 4.6). Finally the cornea

166   Visible and Invisible

FIGURE 4.1 The ocelli of a spider

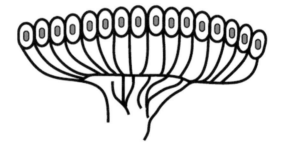

FIGURE 4.2 Primitive eyes were simple light receptors

and the iris were formed as well as a second liquid, the aqueous humor, between the cornea and crystalline lens (Fig. 4.7).

The evolution of various animal organisms (which took multiple independent paths), generated eyes of various kinds. For example, many birds have particularly acute vision, able to perceive even ultraviolet. Insects have eyes composed of numerous autonomous seeing systems. Many fish have eyes on the sides of the head that operate independently; some animals have additional detection organs, and so on.

Evolutionary factors, primarily the ability to recognize food sources, strongly affected the mechanism of color perception.

FIGURE 4.3 Later, the eyes sank to form a cavity

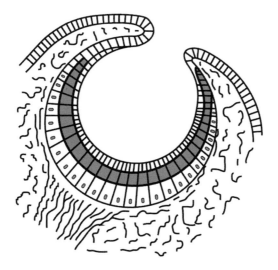

FIGURE 4.4 The formation of the camera obscura

The monochrome eye of some animals, such as seals or whales, simply distinguishes the amount of light, regardless of color, and produces a vision in black and white. Most animals have sight that analyzes the simplest difference of frequencies, selecting a pair of

168   Visible and Invisible

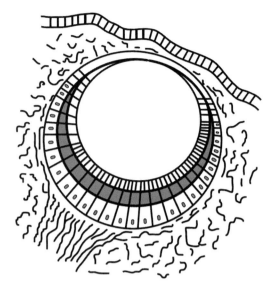

FIGURE 4.5 The closed cavity with vitreous humor inside

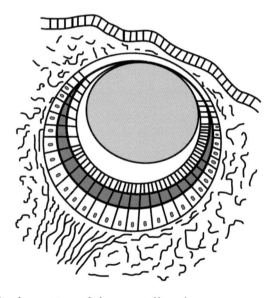

FIGURE 4.6 The formation of the crystalline lens

colors, e.g., blue (high frequency) and yellow (low frequency). Humans and some higher primates, such as chimpanzees and gorillas, developed trichromatic vision with the presence of three types of receptors, sensitive respectively to blue, green, and red.

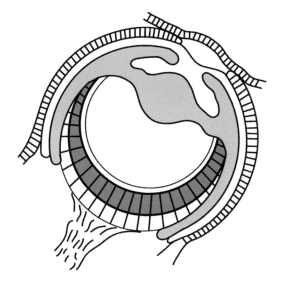

FIGURE 4.7 The formation of the cornea, iris and aqueous humor

Every shade of the color palette may be obtained by combining these three colors. Even bees and hornets distinguish three frequencies; however, they are not sensitive to red but to ultraviolet.

Many birds, several fish species, amphibians, reptiles, arachnids, and insects have four-color vision with sensitivity to ultraviolet light, so they are able to discern colors that are identical to a human being. Some butterflies and some birds, like pigeons, have five types of receptors and are thus believed to possess pentachromatic vision.

Color detection occurs thanks to a type of photoreceptor called a cone. Bichromatic sight is based on two types of cones, trichromatic on three, and so on. Humans and some other animals are also equipped with different types of sensors, called rods, a receptor sensitive to varying shades of gray, and are able to detect very weak light.

The Human Eye

Natural evolution has provided us with a sophisticated instrument, capable of collecting luminous stimuli to be processed by the brain.

170   Visible and Invisible

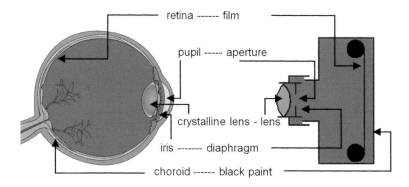

FIGURE 4.8 The similarities between our eye and a camera

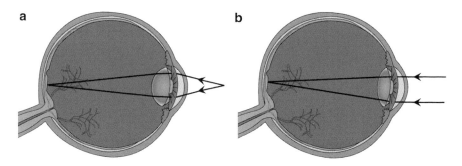

FIGURE 4.9 The crystalline lens is a very special lens, endowed with the capability of changing the curvature of its own surfaces—increasing it if you have to observe nearby objects (**a**), and decreasing it if you have to look at far away objects (**b**)

On first inspection, our eye has a structure similar to that of a camera (Fig. 4.8). Light enters through the pupil, and the image is brought into focus by the crystalline lens onto the retina, upside-down just like in a camera obscura. The image processing is performed by the cerebral cortex, which then takes account of this reversal. The iris determines the opening of the pupil according to the amount of light present. The inner wall of the eye is covered with a black layer, the choroid, which absorbs luminous rays, eliminating those reflected. Therefore the light does not come out from the pupil, which appears black. The crystalline lens is a very special lens that ensures the focus of the image by changing its own shape (Fig. 4.9).

In the retina, there are two types of photoreceptors: cones and rods. The cones are sensitive to color and to daylight, while rods operate in low light conditions. The retina sends the information provided by cones and rods to the brain through the optic nerve.

In the human eye there are three types of cones that allow for trichromatic vision. Yet these do not have the same sensitivity. The most efficient peaks are in the yellow–green region, and these are the only cones able to detect orange and red. The second group has half that effectiveness and shows a peak around green, and the third, much less efficient, responds to blue–violet.

As a consequence, the perception of white light is greatest around the yellow–green wavelength. The processing of the signals received from the three cones allows the brain to recognize all colors.

The outer surface of the eye, which surrounds the colored circle of the iris, is called the sclera. In primate mammals it is of the same color as the skin and thus well camouflaged, while that of humans is white, clearly showing the direction of the gaze. Some scientists have studied this evolutionary diversity by associating it with the unique characteristics of humans, while for another animal it may be useful to hide the direction of the gaze, to avoid challenging a rival, or to fool a predator. Human evolution has favored a different development of the sclera. In particular, the advantage achieved with the use of fire and of tools changed our relationship with other animals, making it possible to gaze. Furthermore, humans' social propensity meant eye signals could be used as a communication tool, for asking for help or when hunting in a group.

CONSIDER THIS
At a certain age it becomes impossible to read without glasses. How come?

CONSIDER THIS
What causes red eyes in many photographs taken with the flash? For what reason are the eyes red and not white or another color in these conditions?

172 Visible and Invisible

Color Perception

Our visual system recognizes colors by using the eye and the nervous sys tem.

The human eye perceives only a restricted part of the electromagnetic spectrum, that found in the visible zone, and not all wavelengths are detected in the same way. The diagram (Fig. 4.10) shows the average sensitivity to different colors. This is at its greatest between green and yellow and minimal towards the edges of the spectrum.

Vision and perception of the colors are completed in the brain where information processing takes place. The main perceptual features are:

- The Opponent Process: The visual sensations arising from the pairs green–red, yellow–blue, and white–black, appear in conflict, i.e., they tend to cancel out each other (Fig. 4.11). You cannot perceive green–red or yellow–blue, whereas red–blue (magenta), yellow–red (orange), and green–blue (cyan) are distinguishable.

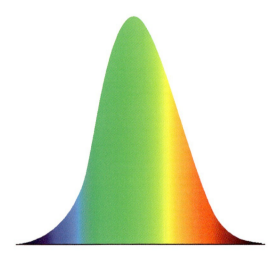

FIGURE 4.10 The sensitivity of our visual system to different colors is highest between *green* and *yellow*

FIGURE 4.11 The antagonist pairs in the opponent process: green–red, yellow–blue and white–black

- Simultaneous Contrast: This is the phenomenon that occurs when the background alters the color of an image. For example, a gray figure assumes different shades if you change the background. Against red the silhouette acquires a green undercurrent, while against green there is a red hint. The same is true for other antagonist pairs, which 'support' each other (Fig. 4.12).
- Color Constancy: If an object is illuminated with various lights, its color does not change. For example, a lemon is yellow at any time of day, even if its color appears slightly falsified in the light of a neon lamp. The eye–brain system considers both the luminous signal coming from the object and the quality of the light that illuminates it. For this property the colors of the bodies remain those familiar to us, even though diffused colors undergo variations.

174   Visible and Invisible

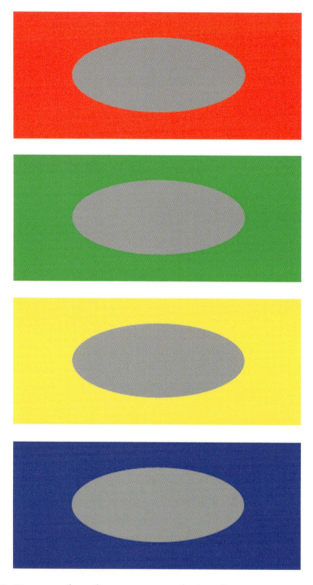

FIGURE 4.12 Due to the phenomenon of simultaneous contrast, a *gray* figure takes on different shades depending on the color of the background

> **EWALD HERING**
> The theory of the opponent process and simultaneous contrast was proposed around 1872 by the German physiologist Ewald Hering (1834–1918), in conflict with the interpretation of vision of an exclusively physical kind, which attributed the perception of color to the activity of only three types of photoreceptors, the cones. These explain the formation of colors but do not describe certain aspects of vision, highlighted by Hering. Today we know that his model is based on precise physical phenomena. The antagonist process is the result of the activity of neurons present in both the retina and in the subsequent visual paths, which are excited by one of the two antagonist colors and inhibited by the other. The perceptual experience of simultaneous contrast is instead associated with the chromatic stimulation of neurons located at the visual cortex level.

Dalton and Defects in Color Perception

Chemistry owes much to the English scientist John Dalton (1766–1844), father of the modern atomic theory.

Dalton was a Quaker, uncompromising in his beliefs and customs, even in his clothing. In his youth, an incident occurred that very much disturbed his mother. The young John, in order to go to a meeting of the congregation, put on a blazing red pair of socks. And yet to him they appeared brown!

The episode led him to carry out a careful and systematic investigation of his visual defect, reaching an initial scientific description in 1794. The publication, entitled *Extraordinary Facts Relating to the Vision of Colors*, exposed the phenomenology of what was to become known as Daltonism (color blindness in a broader typology), a visual defect by which only a few colors can be seen. Dalton was able to recognize the blue–violet and yellow. As for red, in the same text, he writes: "that part of the image which others call red appears to me little more than a shade or defect of light. After that the orange, yellow and green seem one

176   Visible and Invisible

FIGURE 4.13 The visible spectrum as seen by a person not affected by color blindness

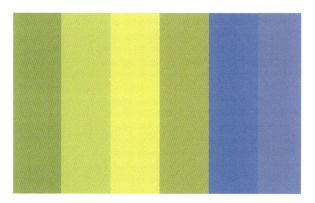

FIGURE 4.14 The visible spectrum as seen by a person affected by deuteranopia (blindness to green)

color which descends pretty uniformly from an intense to a rare yellow, making what I should call different shades of yellow."

Dalton arranged that after his death, his eyes were to be examined in an attempt to discover the cause of the defect. The analysis at the time found no anomalies, but a new study, conducted in 1990, revealed the absence of the pigment that gives sensitivity to green, from which a particular form of color blindness derives, known as deuteranopia (Figs. 4.13 and 4.14).

Deuteranopia is the most common form of color blindness, but it is not the only form. The normal rainbow may undergo various chromatic distortions. Protanopia is characterized by an

FIGURE 4.15 The visible spectrum as seen by a person affected by protanopia (blindness to red)

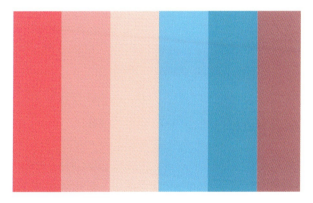

FIGURE 4.16 The visible spectrum as seen by a person affected by tritanopia (blindness to blue)

insensitivity to red, while tritanopia is an insensitivity to blue, and both defects lead to a bichromatic vision (Figs. 4.15 and 4.16). Some individuals suffer instead from achromatopsia, with a visual defect of all three primary colors and thus monochromatic vision.

Individuals affected by deuteranopia, the most common form of color blindness, cannot distinguish red from green. In traffic lights, to avoid any their possible error, it would be sufficient to replace the red–green pair with red–blue.

178 Visible and Invisible

The Grammar of Color

A hue is a pure color, i.e., characterized by a single wavelength within the visible spectrum. The human eye is able to distinguish about 150 light frequencies, corresponding to as many different hues, although the most common are far fewer. For example, the twelve colors of the color wheel are sufficient for many applications.

We can obtain many different colors for every single hue by varying its value (or lightness) or its saturation.

Let us examine first the value, i.e., the sensation produced in an observer by the amount of light reflected, starting from the black–white pair. We may define a sequence by varying the percentage of lightness between these two extremes. For simplicity, we shall consider only seven subdivisions (Fig. 4.17).

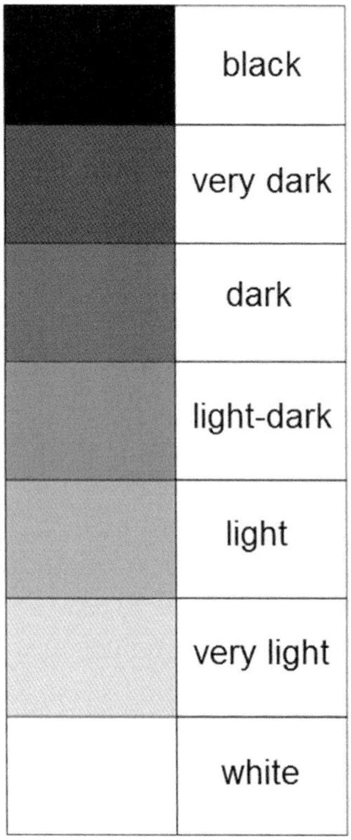

	black
	very dark
	dark
	light-dark
	light
	very light
	white

FIGURE 4.17 By varying the percentage of lightness you can get a scale from *white* to *black*

Light and Life 179

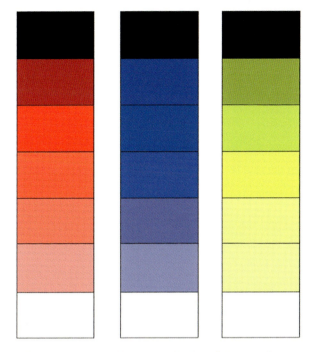

FIGURE 4.18 The sequences that are obtained by applying the scale of Fig. 4.17 to *red, blue,* and *yellow*

Applying this scale to each hue, i.e., by mixing the chosen color with the amount of white/gray/black indicated in the black–white scale given above, we obtain different gradations. In the case of red, blue and yellow, we obtain the sequences shown in Fig. 4.18.

Saturation identifies a different color property—its distance from the pure hue, typical of the color wheel, or its distance from gray. Saturated colors are vivid and intense, while those less saturated appear dull and tend towards gray (Fig. 4.19).

We are now able to determine how many colors there are. Starting from the 150 different hues perceivable by the human eye, and using all possible variations of value and saturation, we may distinguish about 7.5 million different colors!

The various colors can be arranged in a three-dimensional diagram, called an HSL diagram, in which hue, saturation and lightness are represented. A similar diagram is the very useful HSV diagram, which is obtained by replacing lightness with value. This diagram is shown in Fig. 4.20.

180 Visible and Invisible

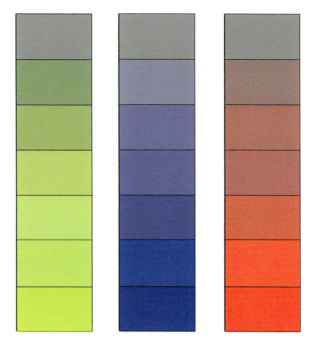

Figure 4.19 Varying saturation leads us to distinguish intense colors as well as other, less vivid ones

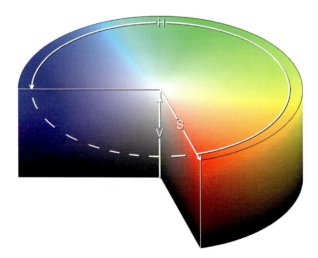

Figure 4.20 An HSV diagram. Hue is measured by an angle centered on the vertical axis, with *red* at 0°, *green* at 120°, and *blue* at 240°. The saturation varies along the radius of the cylinder, the height of which represents the value

The Color Wheel and Harmony

Newton was the first person to roll up the rainbow into a ring, i.e., the color wheel. The German writer and poet Johann Wolfgang von Goethe (1749–1832) did not share Newton's physical explanation of the colors, and in 1810, in his famous book *Zur Farbenlehre* (Theory of Colors) introduced a different interpretation, based on the tensions and harmonies of hues. In this way it was possible to highlight the connections among them, an approach widely used in the art world.

Figure 4.21 shows a color wheel divided into 12 colors, distributed into three groups: primary, secondary, and tertiary.

The three primary colors used by painters are red, blue, and yellow, and are located at the vertices of the triangle pointing upwards. The three secondaries, which are located at the vertices of the triangle directed downwards, are orange, violet, and green. Each of these can be created from two primaries—orange from red and yellow, violet from red and blue, and green from blue and yellow.

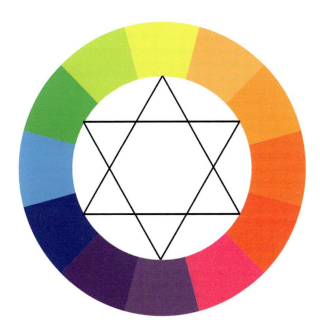

FIGURE 4.21 The color wheel allows us to identify the relationships between the primary, secondary, and tertiary colors

TABLE 4.1 Comparison between the RGB model and the color wheel

	Rgb subtractive mixing	Color wheel
Primary colors	Magenta	Red
	Cyan	Blue
	Yellow	Yellow
Secondary colors	Red	Orange
	Blue	Violet
	Green	Green
Some complementary pairs	Cyan–red	Orange–blue
	Blue–yellow	Violet–yellow
	Magenta–green	Red–green

By combining a primary color with a secondary, a tertiary color is obtained that, in the color circle, is positioned between the pair that originates it. In this way we get six mixtures—yellow–orange, orange–red, red–violet, blue–violet, blue–green, and yellow–green.

The pairs formed by diametrically opposing colors in the chromatic circle, like red and green, are called complementary. Red is primary, while green is composed of yellow and blue. The red–green, therefore, contains the triad of primary colors: red, yellow, and blue. This is true for all complementary pairs.

However, yellow, red, and blue, primary colors of the chromatic circle used by painters, do not coincide with the primary colors of additive mixing (red, green, and blue) or subtractive (magenta, cyan, and yellow) of the RGB model. Consequently, in these two models, also the secondary colors and pairs of complementary colors are different (Table 4.1).

The choice of yellow as a primary color instead of green leads to an approximate complementarity. By mixing two complementary colors from the chromatic circle we only approach an achromatic light, which instead is obtained by mixing yellow and blue, complementary colors in the RGB model.

The color wheel is the primary tool to bring out the relationships between colors, particularly the harmony between adjacent ones and the contrast between opposites.

The analogous colors are those that contain a common color and, when combined, create a natural harmony. They form groups, usually of three elements, close in the circle; for example, blue/blue–green/green, yellow/yellow–orange/orange, red/red–violet/violet, all include analogous colors (Figs. 4.22 and 4.23).

FIGURE 4.22 Examples of analogous colors, formed by three consecutive elements in the chromatic circle

FIGURE 4.23 Can the number of analogous colors be extended? We start by the *blue/blue–green/green* and add colors, following the color wheel—*yellow–green*, *yellow*, and *yellow–orange*. By adding the fourth and even the fifth element, the harmony is preserved although weakened. With the sixth, the yellow–orange, which contains *orange*, complementary to *blue*, you lose the harmony of colors and the contrast appears

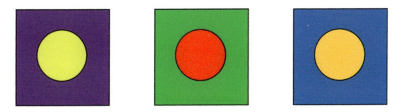

FIGURE 4.24 When a bright color is placed at the center of its less bright complementary, the contrast and complementarity effect is particularly evident

Two complementary colors do not create harmony but are responsible for interesting phenomena. In particular, they present a different brightness. In the pairs yellow–violet, red–green, and orange–blue, the first hue is brighter than the second one.

If the complementary colors are placed side by side, the maximum contrast effect is obtained. Both acquire chromatic strength, mutually reinforcing and increasing the other's brightness (Fig. 4.24).

CONSIDER THIS

The French painters Paul Signac (1863–1935) and Georges Seurat (1859–1891) were sophisticated investigators of color theory. Many of their paintings, produced by putting distinct points of pure color on canvas, were based on the assumption that the vision of the observer could mix the different chromatic tones on the retina (pointillism). In your opinion, is this hypothesis based on additive or subtractive mixing? (Figs. 4.25 and 4.26).

FIGURE 4.25 Femmes au puits, oil on canvas by Paul Signac, 1892

FIGURE 4.26 A detail from Signac's work allows us to appreciate the spirit of pointillism

Deceiving the Eye

When we look at an image, such as a room with a table and chairs, what is it that reaches our eyes? A collection of lines, surfaces, and colors. Our brain has learned to group them in the 'correct' way, giving us the result we are accustomed to. We are able to retain the image of a detail even if it is moved or if it is partly hidden. The lines of a chair are ordered in the right way even if it is behind a table. The brain uses rules selected over millennia and that work in the majority of cases, particularly in circumstances that arise on a daily basis, but we can find many unusual conditions in which its interpretation fails.

If the observed objects are overlapping, their contours are partly hidden. Our brains have learned to reconstruct the forms by imagining the connections that are not visible, and this can lead us into error.

Besides a reconstruction of the shapes, our perceptive system has learned to distinguish one object from a background, but even in this case, anomalous situations may mislead it, as shown in Figs. 4.27, 4.28, and 4.29.

186   Visible and Invisible

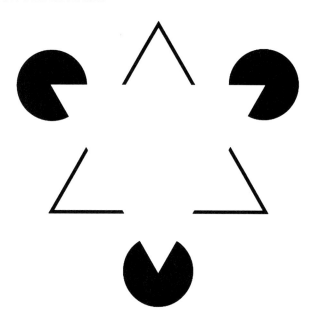

FIGURE 4.27 In the drawing we see a white triangle that is actually nonexistent, but the connection between the different shapes creates an illusory figure

FIGURE 4.28 Even more complex representations, such as those proposed by the American psychologist Joseph Jastrow (1863–1944), can be misleading. Faced with an image that exhibits not one but two usual shapes, which should we choose? Our visual mechanism does not know how to behave. Sometimes it interprets the above picture as the head of a duck, other times as that of a rabbit

Light and Life 187

FIGURE 4.29 Is this a vase or the profiles of two faces?

CONSIDER THIS

In the image in Fig. 4.30, you see two cube faces parallel to the page, but you cannot figure out which one is the closest to the reader. Is that true? The two segments of Fig. 4.31 are of different lengths. Is that true?

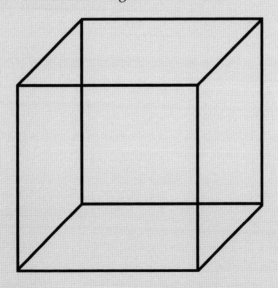

FIGURE 4.30 Which of the two cube faces is the closer?

188 Visible and Invisible

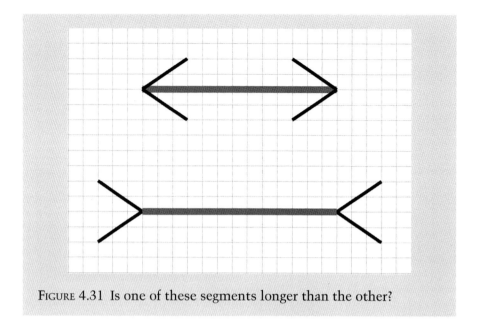

FIGURE 4.31 Is one of these segments longer than the other?

Mirages

In mirages, not all the light rays that reach the eye follow a straight line. Since our brain (or a camera) processes all the signals it receives without distinguishing between curved or straight paths, the vision is distorted. A typical mirage is one that may appear in the desert, with the emergence of a lake full of water. This is not about hallucinations. Mirages are not blunders, but natural processes documented by photographic images and explainable in scientific terms (Fig. 4.32).

Light, in fact, does not always travel in a straight line, but can bend due to refraction. If luminous rays pass through the boundary between two homogeneous media, at the point of interface between the two we have only one deviation, but what happens when light travels through a non-homogeneous substance such as the atmosphere?

In deserts, the air in contact with the ground is hot, thus less dense. Just a little higher, however, it is cooler and denser; therefore, the light that comes from the upper layers passes through

FIGURE 4.32 The appearance of a 'lake' in the middle of a desert

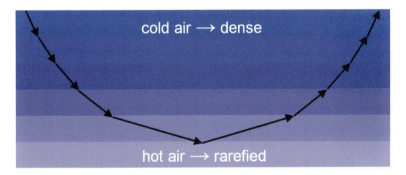

FIGURE 4.33 In the desert the light coming from above follows a curvilinear trajectory, because it passes through layers of increasingly rarefied air

ever more rarefied layers, bending continuously and thus plotting a curvilinear trajectory, as shown in Fig. 4.33.

When our eyes look down, they also receive the luminous rays coming from the sky, which are interpreted as if coming from the ground, i.e., as if they were straight. The result is that portions of the sky appear on the ground, and the observer may interpret the ensuing image as a pool of water (Fig. 4.34).

Mirages are not the only optical phenomenon produced by the curvilinear trajectory that light sometimes takes. The special lights of sunsets are also due to the same cause.

FIGURE 4.34 Our eye interprets the refracted signals as straight, so a part of the sky appears to us as if it were on the ground

CONSIDER THIS
Mirage are not found exclusively in deserts and can also be seen in green and populated regions. On hot summer days, we sometimes see a strange effect on road surfaces. They seem wet and are covered in puddles. The impression disappears as soon as we draw closer, and in fact the asphalt is perfectly dry. Why? The road surface is very hot and heats the air to the point that the image we receive comes partly from the sky. Like in the desert, we have the sensation of a distant fluid. The light appears to shimmer due to air turbulence that locally modifies the refraction process. Can that be true?

FATA MORGANA
Ancient legends tell of the half-sister of King Arthur, Fata Morgana (Morgan le Fay), who according to certain traditions, had the power to build castles in the air. This is the reason why one of the most spectacular mirages carries her name. Such a phenomenon is more complex than those previously described, because is caused by an irregular distribution of the air density. This feature can transform the horizon in a vertical wall with structures that rapidly change their shape.

On August 14, 1643, Father Angelucci was on the Strait of Messina, Reggio Calabria, and described what he observed as follows: "The sea that washes the Sicilian shore swelled up, and became, for ten miles in length, like a chain of dark mountains; while the waters near our Calabrian coast grew quite smooth, and in an instant appeared as one clear polished mirror, reclining against the aforesaid ridge. On this glass was depicted, in chiaroscuro, a string of several thousands of pilasters, all equal in altitude, distance, and degree of light and shade. In a moment they lost half their height, and bent into arcades, like Roman aqueducts. A long cornice was next formed on the top, and above it rose castles innumerable, all perfectly alike. These soon split into towers, which were shortly after lost in colonnades, then windows, and at last ended in pines, cypresses, and other trees, even and similar."

3D Images and the Movies

Natural evolution has endowed us with two eyes that, by observing the same object, are able to estimate its distance and assess its depth. This is how we see in three dimensions.

Unfortunately, the systems that we use to store and transmit images, from photographs to television and movie screens, project the objects in two dimensions, thereby losing their depth. Various techniques, such as perspective, are employed to fill this shortcoming, but the shapes remain flat. Is it possible to store and project three-dimensional images?

Certainly, as long as they are not created for only one eye but for two! Two independent figures must be stored, similar but slightly staggered, the first reserved for the right eye, the second to the left (Fig. 4.35). When you look at the two figures simultaneously, each with the appropriate eye, you regain three-dimensional vision, obviously within the limits of the depth with which they were created. This technique is called stereoscopy.

Modern stereoscopy was born thanks to the English physicist Charles Wheatstone (1802–1875) who, in 1832, perfected an optical instrument, the stereoscope (Fig. 4.36) that was able to obtain a three-dimensional effect by using a pair of two-dimensional images.

192 Visible and Invisible

FIGURE 4.35 Two photographs, taken in 1901, using the technique of stereoscopy

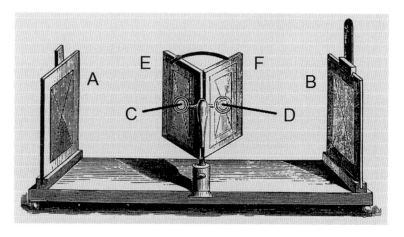

FIGURE 4.36 The mirror stereoscope, developed by Charles Wheatstone in 1832. In the device, the two figures are placed on the side panels A and B. The observer uses the opening C for the *left* eye and the opening D for the *right* one, acquiring the images through mirrors E and F

The same principle is used today for viewing images or movies. For example, in Fig. 4.37, two independent images, rendered using complementary colors, are overlapped and staggered according to the principles of stereoscopy.

Light and Life 193

FIGURE 4.37 Anaglyphic conversion of the two images shown in Fig. 4.35

Normally the pairs of complementary colors used are red–orange (left eye) and blue–green (right eye). An image of this type is called anaglyph. Stereoscopic vision is achieved by wearing glasses equipped with color filters; each eye sees only one color, separating the overlapping images.

CONSIDER THIS
Not all animals have the same vision that humans are equipped with. In general, predators have a similar binocular vision to ours, with eyes facing forward. This visual field is the best suited to their role as hunters. To be able to catch their prey, they must assess the distance with accuracy. Prey, on the other hand, have developed a visual field

194   Visible and Invisible

as broad as possible, which allows them to identify a predator as soon as possible. Their monocular vision, with eyes that receive lateral and disjointed information, is very effective as a lookout tool. So, why do humans, who get their nutrition mainly from farming, have binocular vision? (Fig. 4.38)

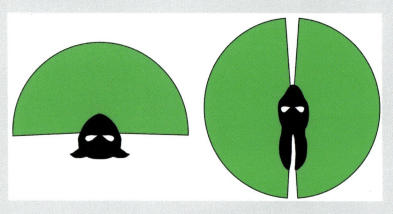

FIGURE 4.38 The visual field of a predator (the cat) and of a prey (the rabbit)

SEE IT IN DEPTH

How can two eyes, looking at the same objects from two slightly different positions, calculate their distance? It's all thanks to the brain, which processes the two images and reconstructs the depth. Let us consider, for example, the gray sphere in Fig. 4.39. Based on the point of observation, it appears to be located differently against the background; one eye places it to the right, the other to the left.

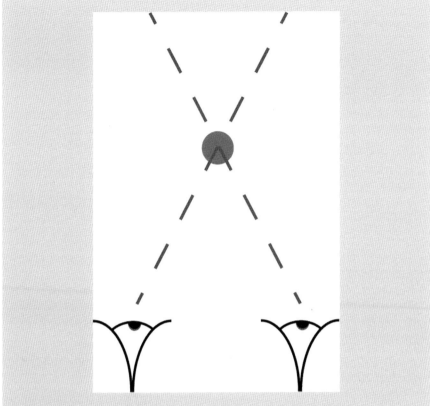

FIGURE 4.39 The *gray sphere* appears more to the *left* when viewed with the *right* eye, and more to the *right* when viewed with the *left*

If the sphere is very close to the observer, this difference is significant, while it is less significant if the distance is greater.

This is the mechanism that enables our brain to perceive objects in three dimensions. Astronomers use this same phenomenon, called parallax, to determine the distances of the celestial bodies.

> **AN ADVANTAGE OF STRABISMUS**
> Binocular vision requires perfect alignment, which in strabismus is missing: this visual defect may impede the evaluation of depth.
>
> While for most squinters this characteristic entails a disadvantage, for painters, committed to represent three-dimensional images on a flat canvas, it can help. It has been discovered that many great painters, including Il Guercino, Rembrandt, Picasso, Klimt, Chagall, Man Ray, Hopper and Lichtenstein had a definite squint.

Bird Sight

Over millions of years, evolution has created a great multitude of species, from which humankind stands out. The eye played a key role in the development of humankind, even though our visual system does not excel. Birds, for example, have developed a better visual apparatus.

One feature of a bird's sight, shared with other animals, is its having four types of cones, while our sight uses three. The fourth cone is sensitive to ultraviolet light. What are the implications of this difference?

The presence of three receptors in the human eye implies that all detectable colors may be obtained by mixing three hues, namely red, green, and blue. The four types of cones found in birds provide greater variety. For example two forms, which we see as being identical in color, may not appear in the same way to them, given the differences in the distribution of ultraviolet light. So just as our eyes, by adding different amounts of red and green, perceive different shades of yellow, a bird's eye detects different colors when it receives red light and ultraviolet light in varying proportions. Consequently, these animals are able to see hues that we cannot even imagine (Fig. 4.40).

What advantages does such chromatic richness provide for birds? The males are more colorful than females, and therefore it plays a role in partner selection.

Light and Life 197

FIGURE 4.40 The beautiful coloring of the plumage of a male Painted Bunting has to appear even more spectacular to the eyes of a female when she observes a potential partner. By studying the light diffused by the plumage it was discovered that the ultraviolet component varies from area to area

In the specific case of the Blue Grosbeak, it has been discovered that the males that reflect the most ultraviolet light are bigger, control the best territory, and feed their offspring with greater frequency. Furthermore, a bird's sight is useful when seeking fruits and berries that diffuse rays invisible to us.

Sensitivity to ultraviolet light is not the only characteristic of bird sight. The eyes of birds of prey, for example, are proportionately large, so as to give them that expression of irresistible beauty and a typical look somewhere between proud and wild (Fig. 4.41). Their large eyes provide them with very powerful sight, capable of operating like a telephoto lens, enlarging every detail. In these animals, the number of photoreceptors is high, five times that found in the human visual apparatus. Thanks to these properties, the images they observe are magnified and provide very high definition.

198 Visible and Invisible

FIGURE 4.41 A wonderful close-up of a *white*-headed sea eagle, showing its large eyes

Insect Sight

Insects constitute a very wide group, including more than a million species. Their sight is mainly based on compound eyes, i.e., formed by hundreds, sometimes thousands, of elementary sight units, called the ommatidia. The visual system with the greatest number of ommatidia is found in species that feed in flight, such as dragonflies (Fig. 4.42).

Each ommatidium acts as a rudimentary eye, directed in its own autonomous direction, with a crystalline lens that focuses light onto photosensitive cells. These generate a signal that, through the optic nerve, transmits a portion of the image to the brain, where the overall shape is finally obtained by piecing together the various contributions, like in a mosaic (Fig. 4.43).

The compound eye looks like a less efficient visual tool than that of more evolved animals, including humans, but in return, it is particularly suitable to detecting the trajectory of moving bodies. The movements, in fact, impact the various ommatidia in sequence, so that the insect is able to grasp them with precision.

Light and Life 199

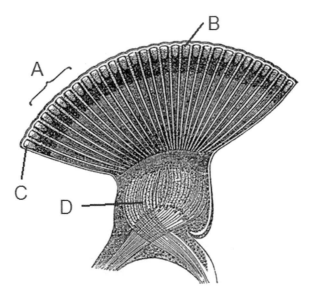

FIGURE 4.42 Scheme of the compound eye of a dragonfly. Wach elementary unit (ommatidium) possesses a cornea, and the eye is covered by a set of corneal facets. *A*, *B*, and *C* indicate the crystalline lenses, while *D* shows the beams of the optic nerves

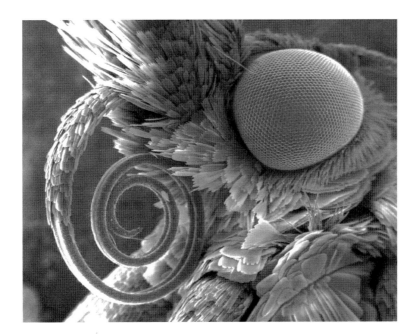

FIGURE 4.43 Head of a moth examined under a scanning electron microscope. The compound eye is about 0.8 mm in diameter. The microscopic elements that form it are the ommatidia

Color perception depends on the species. Many insects have a bichromatic vision, with only two types of cones. Bees and hornets, on the other hand, distinguish three frequencies corresponding to yellow, blue, and ultraviolet.

Fish Sight

Fish sight depends essentially on the characteristics of the water in which the fish live. The underwater world has different lighting conditions, with a continuous range of bright environments, from the intense light of the surface to the darkness of the deep waters.

When the water is clear, such as in tropical oceans or in clean lakes, the colors violet and red are absorbed in the first 25 m, while blue dominates the underwater environment up to more than 75 m. At 100 m it is no longer possible to distinguish any color.

If the water is not clear, like in many lakes, rivers, and swamps, light absorption is rapid. Even blue is absorbed, and a yellow–green color emerges that does not exceed 30 m in depth.

So how did fish eyes evolve in these conditions?

The numerous fishes living in shallow water have trichromatic vision, with three types of cones. For species living at greater depths, visual perception is at first dichromatic, then monochromatic. Furthermore, the size of the eyeball can extend with increasing depth, in order to collect a greater quantity of light. The importance of rods, cells able to detect dim light, also increases to the point that in some fish, these are the only photoreceptors.

Like for some terrestrial animals, the view in the underwater world may be sensitive to infrared or ultraviolet light. The image focus in fish is unique. The crystalline lens does not deform but moves back and forth (Fig. 4.44).

At depths greater than 200 m, we enter the domain of the abyssal species, with very specific forms, suited to an environment characterized by lack of sunlight, still water, a constant temperature (between 0 °C and 5 °C), and intense pressure.

In this habitat, some marine animals are able to produce light through special organs distributed on the body called photophores, and their eyes are equipped with photoreceptors suited to detecting the light produced by these. Lastly, in completely dark environments, there are fish that do not require sight and are therefore blind (Fig. 4.45).

FIGURE 4.44 The dusky grouper is a fish commonly found in the Mediterranean, from 10–150 m deep. It can rotate its eyes independently of one another; for example, it can look forward with one eye while the other scans upward

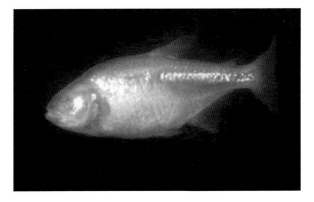

FIGURE 4.45 The blind cave tetra is a small freshwater fish characterized by the absence of eyes, atrophied during evolution until they disappear. The visual organ in this case is useless because the animal lives in totally dark caves

Most fish have eyes on both sides of their head. This gives them a wide field of vision and lets them identify approaching danger. Instead, for those species that lie flat on the seabed to hide from predators, they sometimes have both eyeballs on the same side.

Many flatfish, like sole, are born with an eyeball on each side. As they grow, however, a process of metamorphosis takes place in which the one eye moves towards the side that normally faces upward. Adults have both eyes on the same side.

The Sight of Some Snakes

Crotalinae (a subfamily of vipers that includes rattlesnakes), pythons, and some boas can strike with precision even at night, when their prey is hidden in darkness. How is it possible?

These species can detect the infrared radiation emitted by warm-blooded animals, the body temperature of which is greater than that of the environment. The recognition takes place by means of organs that detect the temperature increase caused by such rays (Fig. 4.46).

FIGURE 4.46 The head of a rattlesnake. The *black arrow* indicates a nostril while the *white* one identifies the opening of one of the two dimples

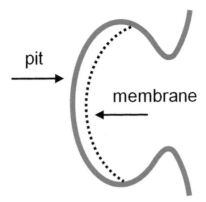

FIGURE 4.47 The opening of the dimple is fairly small compared to the size of the cavity

These organs are nothing more than small cavities inside which a very thin heat-sensitive membrane is located, suspended in the hollow and connected to the brain through a branch of the trigeminal nerve. When the temperature of the membrane changes (the minimum detectable variation is 0.003 °C) a signal is sent (Fig. 4.47).

The infrared radiation that enters the opening is projected onto the membrane, like in a camera obscura. This allows the snake to identify the position of the warm body emitting infrared rays with sufficient precision, thus providing these animals with an extra sense.

Only cold-blooded animals, such as snakes, may take advantage of such a faculty. If a warm-blooded being were equipped with a device like this, it would fail to detect the presence of other animals because it would be 'dazzled' by its own body heat.

A thermal image is certainly less accurate and detailed than a visual one, but it has the advantage of being able to detect a warm body in the dark or when hidden. Therefore, it is complementary to optical vision.

Figures 4.48, 4.49 and 4.50 refer to the same object observed during the day: a hand, hidden behind a bush. In Fig. 4.48 the form is detected with infrared light, Fig. 4.49 shows the same shape seen through using visible light, while Fig. 4.50 shows the overlapping of the two, what is received by combining the two signals.

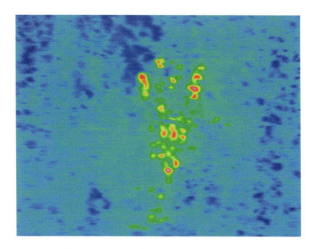

FIGURE 4.48 A hand hidden behind a bush detected with infrared rays

FIGURE 4.49 The visible-light image of the same hand

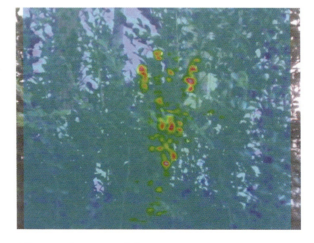

FIGURE 4.50 The overlapping of the two previous images

Colors and the Survival of the Species

In the animal world, color is an important protagonist of natural selection, particularly in survival strategies. It plays a significant role in both sustenance (obtaining food) and safety (escape predators).

The camouflage of predators allows them to approach potential prey without being seen. The tiger, at the top of the food chain, having no predators in the animal environment, possesses a sophisticated mimetic system. It lives and hunts among bamboo plants, and the background in which it moves consists of yellow trunks against dark spaces—the same colors as its fur (Fig. 4.51).

Also those animals that are preyed upon have developed camouflage strategies. A well-known case is that of chameleons, a family of reptiles that in fact includes numerous species. The fur of many animals never changes, while for others the color can alter with the passing of the seasons. The ermine, for example, has a brown coat during the summer that turns white as snow in winter.

Finally, there is a third group of animals, including chameleons, which are able to change their appearance in a very short time, for particular moods or camouflage requirements. In chameleons the color change is due to specialized cells that lie

FIGURE 4.51 A tiger camouflaged among the bamboo plants

206   Visible and Invisible

FIGURE 4.52 The survival strategy of the *Bradypodion atromontanum* chameleon. The image shows the joint effect of color and position

beneath their transparent skin and which are activated directly by the nervous system, allowing for quick changes, from milliseconds to seconds (Fig. 4.52).

In nature, we may also observe the opposite behavior to camouflage. In many species females and males are different, the latter being larger and displaying more colors.

The evolution theorist, Charles Darwin, analyzed this phenomenon, identifying it as sexual selection. According to the English naturalist, individuals with particular characteristics, such as large size and lively coloring, are more likely to be successful when competing with other males to conquer the female and thus procreate (Fig. 4.53).

Bioluminescence

Animals such as fireflies or jellyfish, some fungi, algae, and even some bacteria can emit light. This property is useful both as a defensive tool and to facilitate encounters for mating purposes (Fig. 4.54). Some fish that live in the deep sea, enshrouded in absolute darkness, have organs able illuminate the environment, useful to attract prey.

Light and Life 207

FIGURE 4.53 The beautiful tail of a male peacock. In the female, such colors and shapes are less evident. In this way it is able to hide itself more easily, increasing its chances of going unnoticed by predators. On the other hand, the male characteristics, useful for finding a partner, are a disadvantage in terms of natural selection. For the male, the need to seduce a mate prevails over the survival of the individual

FIGURE 4.54 In the mating season, fireflies use bioluminescence to attract a partner

208 Visible and Invisible

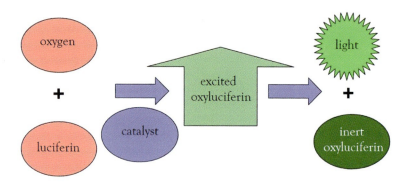

FIGURE 4.55 Scheme of the chemical reaction that produces bioluminescence

What produces this light? The most common mechanism is that of a chemical reaction between two substances in the presence of a catalyst, which accelerates the process. The first substance required is found in the environment—oxygen—while the second is produced by the bioluminescent organisms and is called luciferin, meaning "light-bringer." The catalyst is an enzyme called luciferase, and the reaction may take place in special organs called photophores (Fig. 4.55).

The chemical reaction leads to the creation of a product (oxyluciferin) in an excited state, which decays by emitting photons. The rays emitted may have different colors. About 95 % of the energy appears in form of light, which accordingly is cold.

Bioluminescence is widespread in the marine environment, constituting a primary form of illumination in the oceans. Sailors' legends are full of episodes in which the ocean at night produces an intense glow, which may be compared to a stack of clouds or a snowfield (Fig. 4.56).

This phenomenon has also attracted famous writers such as Jules Verne who, in *Twenty Thousand Leagues Under the Sea*, writes:

> [T]he sea looked as if it was illuminated from below. There could be no mistake, for this was no ordinary phosphorescence. Several fathoms below the surface, the monster gave forth a very strong, inexplicable light, as described in the reports of several captains.

Light and Life 209

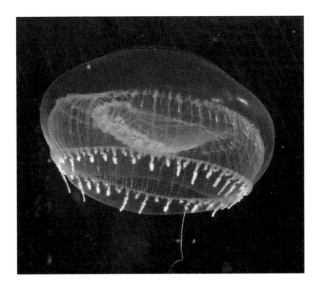

FIGURE 4.56 The crystal jellyfish is a bioluminescent marine organism

This fantastic irradiation must have been produced by some tremendously powerful illuminating agent. The radiant area on the surface formed a huge, highly elongated ellipse, whose centre was a burning condensed focus of unbearable intensity, but which gradually faded further away from the centre. "It's just a mass of phosphorescent organisms!" cried one of the officers..... (Fig. 4.57)

CONSIDER THIS

Jules Verne describes the light emitted by marine organisms as 'phosphorescent'; the expression is also used by the scientist Charles Darwin who, in his book *The Voyage of the Beagle*, thus describes a zoophyte, i.e., a marine animal, such as the sea anemone, which resembles a plant: "Having kept a large tuft of it in a basin of salt water, when it was dark I found that as often as I rubbed any part of a branch, the whole became strongly phosphorescent with a green light. I do not think I ever saw any object more beautifully so." Can that be true?

210   Visible and Invisible

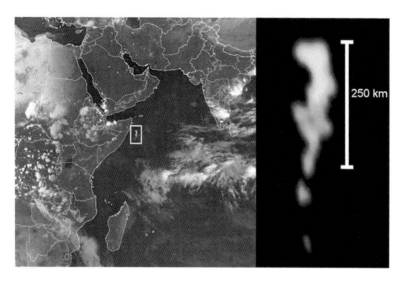

FIGURE 4.57 An evocative image, which recalls the description of Jules Verne, was provided by a satellite on January 25, 1995. The photo shows a large area of milky sea at about 280 km from the coast of Somalia. The effect is caused by a large amount of bioluminescent bacteria present in the region

Photosynthesis

In order to survive, a predator feeds on meat, by hunting prey that may be predators of lower rank or herbivores. At the end of the animal food chain there is always a herbivore. The whole animal world depends on the vegetable world, without which herbivores could not feed, meaning that predators would be left without prey. But if the animal world, directly or indirectly, feeds on plants, what do plants feed on?

They feed mainly on water, carbon dioxide and ... light. Plants absorb luminous energy and activate a process, photosynthesis, that primarily produces glucose, their own food, and a waste product essential to life—oxygen (Fig. 4.58).

Light is essential because it provides the energy needed to produce complex molecules from simpler ones. Light rays are absorbed by the leaf through a green photosensitive pigment,

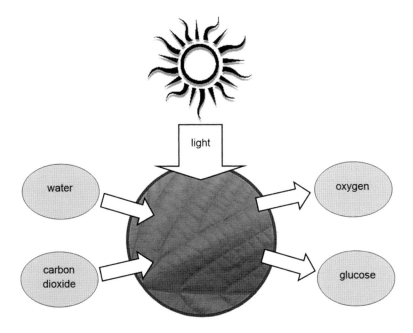

FIGURE 4.58 Leaves absorb light, water, and carbon dioxide that they transform, through complex chemical reactions, into glucose and oxygen. This process consists of a re-mixing of three types of atoms: hydrogen, oxygen, and carbon. Water (hydrogen + oxygen) and carbon dioxide (carbon + oxygen) form glucose (hydrogen + oxygen + carbon) and pure oxygen. Glucose is one of the most important carbohydrates and is used as a source of energy by plants and animals alike

chlorophyll, which thus reaches an excited state for a very short time (a few billionths of a second), during which it promotes the chemical reaction that transforms carbon dioxide and water into glucose and oxygen. The unused energy is re-emitted through a weak fluorescence, while oxygen is released into the atmosphere in the place of carbon dioxide, which is absorbed. Animals, including human beings, live with the reverse process. They inhale oxygen and exhale carbon dioxide. Plants thus enable the preservation of an atmosphere suitable for life on our planet (Fig. 4.59).

🔍 THE OXYGEN CATASTROPHE

The earliest forms of life evolved on Earth about 3.5 billion years ago into structures that did not require the presence of oxygen. The great development of cyanobacteria generated increasing amounts of this molecule, first absorbed by rocks, then issued directly into the air. As a result, Earth's atmosphere has changed radically, becoming rich in this new substance, toxic to many organisms existing at that time. About 2,400 million years ago, the 'oxygen catastrophe' took place, the first major mass extinction in Earth's history. Since then, life on our planet has evolved with organisms that use oxygen.

🔍 CONSIDER THIS

Evolution has developed an effective system for the exploitation of solar energy, stored in plants through photosynthesis. Even today this remains the most efficient method. Can that be true?

FIGURE 4.59 In terrestrial plants, the release of oxygen takes place into the air and is therefore unnoticeable, while in aquatic plants it causes visible bubbles

The possibility of using light to produce food is not exclusive to the plant kingdom, but is also characteristic of some bacteria. Cyanobacteria behave like plants and release oxygen, while other groups of these microscopic organisms give off various other elements, such as sulfur.

The Colors of Leaves

The 'factory' that produces nutrition for the vegetable world is the leaf, which is normally flat and thin, in order to absorb as much light as possible. In summer it is green, whereas in fall it may be yellow, orange, red, or purple. What is the cause of this change?

The green of the leaf is directly related to its function—to carry out photosynthesis. The longest wavelengths (red and orange) and the shortest ones (blue and violet) of the visible spectrum are absorbed by chlorophyll, which instead reflects the central wavelengths (yellow and green), those responsible for the most widespread coloring of the leaves (Fig. 4.60).

In winter there is not enough light and water for photosynthesis, and so in most plants this activity is suspended. Without nutrients, deciduous plants survive using the substances accumulated

FIGURE 4.60 Typically the leaves of plants are *green*

214   Visible and Invisible

FIGURE 4.61 In many trees, such as maples, the termination of photosynthesis traps glucose in the leaves, a substance which then takes on a *red* color

over the summer, during which their leaves produce glucose in abundance. The excess part is converted into starch and conserved. Throughout the cold months, these plants cease food production and eliminate their green chlorophyll. Consequently, at the time of the disappearance of bright green, yellow and orange colors appear, due to the presence of carotene, hues already present in the leaves that are no longer hidden. The color red in some leaves is instead due to the presence of glucose trapped in them (Fig. 4.61).

The Colors of Other Worlds

People have long wondered whether there are other planets endowed with life forms, and there are various research projects underway to try to locate them. The discovery of life forms does not necessarily imply the existence of civilization. Our planet has been dominated by humans only over recent centuries. Assuming that alien life exists, it is likely to consist of elementary organisms, similar to those that have inhabited Earth for millions of years.

FIGURE 4.62 Two examples of what an alien environment might look like, according to NASA

If a planet is home to living beings, it is conceivable that photosynthesis is also present, given that it has been so successful in promoting the development of life on Earth. Natural selection promotes projects with positive results.

Scientists believe the development of organisms capable of creating their own food is possible by using the light of a star, and think that the dominant color of alien vegetation might be red, blue, or even black.

What determines the prevailing tint in the vegetation of a planet? In the first place, the type of light it receives from the mother star.

The characteristics of the radiation depend on the temperature of the star. If it is very hot, then it has too short an existence to allow for the evolution of complex life forms. Research therefore focuses around cooler stars, such as those that are yellow–white, yellow (like our Sun), yellow–orange, and red. Each of these is surrounded by a habitable zone, a range of orbits in which the temperature of the planets allow for the presence of liquid water.

Around bodies hotter than the Sun (yellow–white stars), blue radiation is intense. This being very energetic, it is conceivable that the evolution of life forms similar to ours might employ pigments specialized only in absorbing blue light, enough for photosynthesis. The other colors would be diffused, and vegetation would therefore be red, yellow, or green (Fig. 4.62).

For those planets close to hot stars, a second possibility could be considered. The star might send a flow of blue light too intense

FIGURE 4.63 If a hot star emitted a flow of *blue* light too intense to be absorbed by leaves, the vegetation would appear *blue*

to be absorbed by the leaves. These rays would therefore be reflected, and the vegetation would appear blue (Fig. 4.63).

Finally, let us consider the cooler stars, such as the red ones, which emit less radiation than the Sun. In this case it would be necessary to absorb all the light, from the visible to infrared, so the plants of these planets might be black.

Many years ago, astronomers observed a change in the color of the planet Mars, a phenomenon that was thought to have been caused by a seasonal cycle of vegetation. Could there be life on Mars? Today we know that this assumption is false, since there are no plants on Mars, but periodic and violent storms of reddish sand.

The possible presence of life on this planet is at the heart of the various stories and myth surrounding possible life on Mars. One of the most famous science fiction books is *The War of the Worlds*, written in 1898 by the Englishman Herbert Wells (1866–1946). In this book the Red Planet is described in these terms: "Apparently the vegetable kingdom in Mars, instead of having green for a dominant color, is of a vivid blood-red tint."

5. Light Techniques

> "Creativity is above all the ability to continuously ask questions."
>
> Piero Angela, writer and science journalist

Solar Clocks and Sundials

Since ancient times, people have exploited the position of celestial bodies to determine the passing of time. Obelisks are examples of structures able to determine both the time and the day. The shadow cast on the ground moves according to the different positions of the Sun, providing a direct indication of the flow of time (Fig. 5.1).

The possibility to tell the time in this way is not restricted to the monumental obelisks. It can be achieved by fixing a pole to the ground. This device is called gnomon.

How is it possible to know the hour and the day from the examination of the trajectory of the shadow? This moves throughout the day and during the winter reaches its maximum length.

To understand how this calendar-clock works it is necessary to consider the relative motion of the Sun compared to Earth. From our point of view, the star follows a trajectory from east to west. Its motion is slow, and at noon the shadow is aligned along the north-south axis (Fig. 5.2). With the passing of the days, the path is similar, but it appears slightly rotated, moving between two extreme orbits corresponding to the summer solstice (June 20 or 21) and the winter solstice (December 21 or 22).

218 Visible and Invisible

FIGURE 5.1 An artistic representation of a great ancient solar clock, the Horologium Augusti, built on the Campus Martius in Rome by the Emperor Augustus in 9 B.C. around an obelisk brought back from Egypt following a military victory

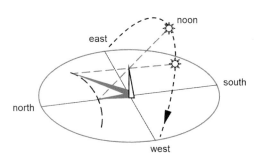

FIGURE 5.2 The shadow of the gnomon moves according to the position of the Sun: at noon it is aligned along the north-south line

Also the length of the night reaches its extreme values in these two dates—its minimum in the former and its maximum in the latter. Instead daytime and nighttime hours are the same length at the spring equinox, on March 21, and the autumnal equinox, on September 21.

Light Techniques 219

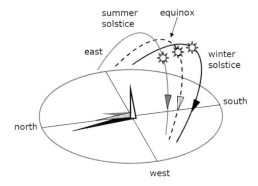

FIGURE 5.3 The shadow projected by a gnomon during the solstices and equinoxes

In Fig. 5.3 the position of shadow on these special days is shown. Shortly after 12 at the summer solstice (path and shadow in red), shortly before 12 in the winter (path and shadow in blue), and at noon during the equinoxes (path and shadow in green). The calculation shown in Figs. 5.4 and 5.5 was carried out assuming that the solar clock were located in Modena, Italy.

A close relative of solar clocks is the sundial, which does not consider the daily trajectory of the Sun but signals only the central moment when the star reaches its maximum height on the horizon. Throughout the year, the noon shadow moves along the meridian line along the north-south axis.

Frequently, the trace of the sundial is not a shadow but a Sun ray that enters through a hole in a low-light environment. Devices of this type are called camera obscura sundials. A well-known one is located in the Basilica of St Petronius in Bologna (Fig. 5.6). Designed by Giovanni Domenico Cassini in 1655, it was used for astronomical and geographical research of great interest. The accuracy of the information it provides is due to its size: the hole from which the solar beam enters is 27 m in height, so the line covered by light in the various days of the year is 67.5 m long.

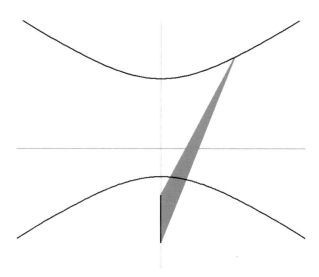

FIGURE 5.4 This drawing shows the line covered by the shadow of the gnomon the day of the winter solstice. The one represented here corresponds to 2 pm. The lower trajectory it is the one covered by the shadow 6 months later, on June 21, while the *gray line* refers to the spring and autumn equinoxes. The *vertical line* that divides the trajectories in half, called the meridian, indicates the north-south axis, with north leading upward. The solar clock in fact also provides a range of geographical and astronomical information

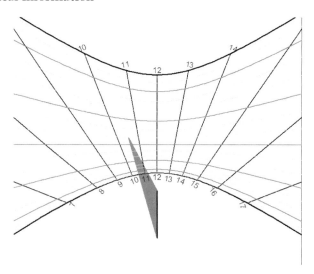

FIGURE 5.5 The path traced by the gnomon in 20 days. Some *lines* are also indicated that start from the gnomon and broaden out. They indicate only the daylight hours, of course, for the clock does not work without sunlight. The *shadow* drawn is that of 10.30 am, on March 10. Figures 5.4 and 5.5 were developed with the Shadows program by François Blateyron, an amateur astronomer and software developer, assuming that the solar clock is in Modena, Italy

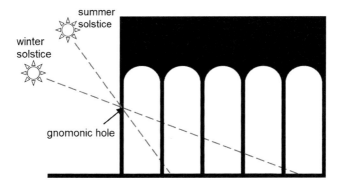

FIGURE 5.6 Outline of the camera obscura sundial that is located in Basilica of St Petronius in Bologna

> **CONSIDER THIS**
> The solar clock marks the hour of the location where it is placed. When the Sun reaches the highest point of its trajectory it is exactly noon, but this indication rarely coincides with that of our clock, synchronized with the radio or television. Why?

Overhead Projectors

There are different types of overhead projectors (OHP). The simplest model consists of a base containing a concave mirror that directs the light of a lamp towards two lenses. The first lies below the working surface while the second makes the rays converge towards a tilted mirror. Adjustments may be made by raising or lowering the second lens, thus altering the image reflected from the mirror onto the screen (Fig. 5.7).

The mirror and the lenses are important. The image projected onto the screen is inverted compared to the object on the working surface (Fig. 5.8).

This transformation passed from the object on the working surface to its image is not the only one achieved with an overhead projector, which turns out to be an interesting means for

222 Visible and Invisible

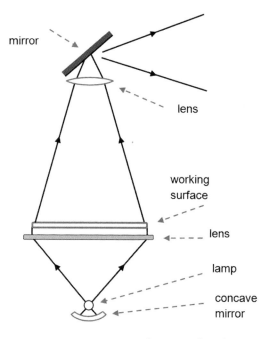

FIGURE 5.7 Schematic representation of an overhead projector

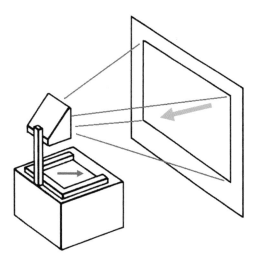

FIGURE 5.8 The image projected by an overhead projector is turned upside-down compared to its original positioning on the working surface

FIGURE 5.9 In projection through an OHP, the vertical dimension of objects is lost

experimenting with the properties of light. For example, the projection loses the vertical dimension, so if you want to preserve the shapes of objects, you need to put them down on the working surface (Fig. 5.9).

The OHP is also very useful for investigating the properties of light propagation, by varying the distance between overhead projector and screen, indicated with a D in Fig. 5.10. When this happens, the image size changes proportionally. If distance D is doubled, the same occurs for the height and the width of the projected image, the area of which consequently quadruples. The linear dimensions are therefore directly proportional to the distance, while the illuminated area is proportional to its square.

The light propagates with an intensity that decreases with the increasing distance covered as the illuminated surface extends. If distance D is doubled, the light intensity is reduced by four times. This result is the basis of various analyses, such as those related to Olbers' paradox.

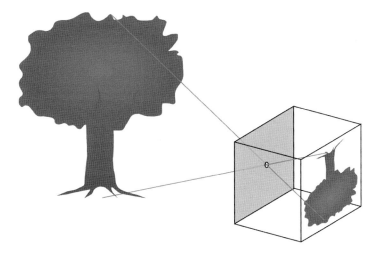

FIGURE 5.11 In a camera obscura the rays, which can enter the opening, project an overturned image onto the opposite wall

The optical instrument described is called a camera obscura and is a device studied and used since ancient times. Aristotle speaks of it in one of his writings, *Problemata*, and it is said that he himself observed a solar eclipse in a small darkened environment. This device, very useful for the study of optics, was later upgraded by inserting a lens in the opening, to better focus the figures. Sometimes a mirror is also used, which sends the image upwards onto a frosted glass that acts like the wall of the box, to unbend the image. This is the principle of the reflex apparatus (Fig. 5.12).

The device in which light enters through a small hole to form an image is not a human invention but a product of natural evolution. The eye operates essentially like a camera obscura.

The possibility of easily obtaining the rigorous planar projection of a landscape profoundly changed the approach to painting in the eighteenth century. Architecture gained an identity of its own and became the subject of artistic works, losing their role as mere background. A new genre, called the veduta, became widespread, oriented towards representing perspective views of cities or landscapes. This type of painting developed mainly in Venice (Fig. 5.13).

FIGURE 5.12 Reflex camera obscura from the early nineteenth century

In pictorial representation, the camera obscura was used not only for the planar projection of three-dimensional views but also to reproduce the wide range of brightness found in nature. It is believed that great artists like Jan Vermeer (1632–1675) made meticulous use of this tool to better define and analyze images.

CONSIDER THIS
The camera obscura is the basis of how cameras work.
In analog equipment, the image is fixed on a photosensitive film, while in digital ones it is stored in an electronic device. If you open up the back of a camera and replace the film with a sheet of paper that serves as a screen, by keeping the shutter open you can see the image turned upside-down on the back of the sheet of paper. Is that true?

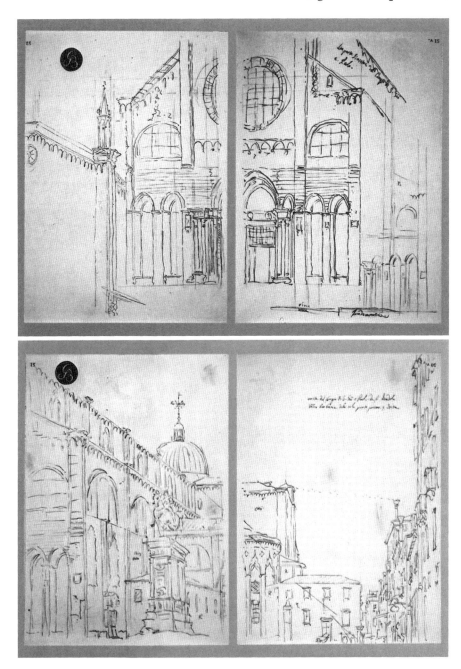

FIGURE 5.13 Study of the Venetian Basilica of Saints John and Paul, by Canaletto (1697–1768) with the help of a camera obscura

Image Manipulation

According to Helmut Gernsheim (1913–1995), an important collector and famous photography historian, this technique defines "the only 'language' understood in all parts of the world, and, bridging all nations and cultures, it links the family of man." Not only is it a universal language, but it is undeniable proof that a thing actually happened. A witness may be confused and remember badly, but the presence of an image clears up any doubt, unambiguously documenting the event. Seeing is believing!

Are we sure this is right? Actually, images can be manipulated and the documentation of the event distorted. For example, look at Fig. 5.14. Although this has always happened, with the advent of the digital age, such opportunities have expanded dramatically. A photograph is no longer undeniable proof. Today it is simple to manipulate a photo because the digital image is nothing more than a collection of numbers.

Let us consider the photo of a well-known painting, *Il Campo di Rialto* by Canaletto (1697–1768), shown in Fig. 5.15. The digital image, like all others, is made up of many elements, so small as to be indistinguishable to the naked eye.

By enlarging a detail repeatedly, such as the lady with the baby to be found in the lower left, we see a grid appear, made up of many small squares, each of which has no internal structure and is characterized just by its coloration (Fig. 5.16). In the RGB model, the primary colors for additive mixing are red, green, and blue, and each color comes from a combination of these. The weight of each one in the mixture is indicated by a number that can vary from 0 to 255. The distinctive feature of each cell is therefore provided by three numbers. The figure, therefore, is nothing but the collection of many constituents, called picture elements (or pixels). Editing a digital photo is therefore simple. Just touch up the color of a part of the small squares that compose it.

Changing the color of a group of pixels not only allows us to modify the overall hue but also to insert new objects. If we color a line of small squares with black, we add a black line to the image. That's why seeing is no longer believing (Fig. 5.17).

Light Techniques 229

FIGURE 5.14 The picture on the *top* is original, the one below is manipulated: the two people in the foreground and their shadows have been eliminated

Seeing in the Micro World

The microscopic environment, from biological cells to the simplest atomic structures, is complex and lively, but it is a reality that our eyes cannot see directly.

230   Visible and Invisible

FIGURE 5.15 A photograph of the famous painting Il Campo di Rialto, by Canaletto

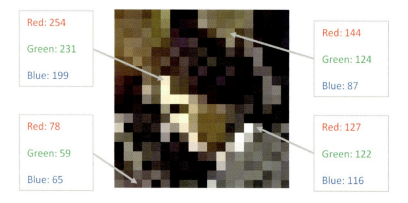

FIGURE 5.16 By enlarging a picture, a grid begins to appear made up of many small squares, called pixels. For each one three numbers are given, which indicate the weight of each of the three RGB colors that create the actual color

How can we examine it? With the aid of the lens, capable of providing the eye with a larger image of the real object. Only one lens, however, is not powerful enough. Systems composed of two or more lenses, or microscopes, must be used.

Light Techniques 231

FIGURE 5.17 These six pictures are taken from a single photograph. Which one shows the true color of the car?

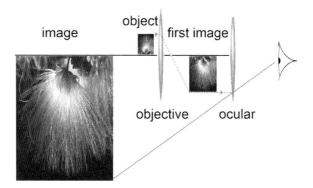

FIGURE 5.18 The image we see under the microscope appears to us magnified because our view interprets all the light signals as if they were rectilinear

The object to be examined, for example the detail of a flower, is placed in front of a first lens, the objective, to form an expanded and inverted image that, before reaching the observer, is enlarged once again by a second lens, called the ocular. At this point, the seeing mechanism interprets the light received by the eye as if it originated from rectilinear rays coming from a figure greatly enlarged and inverted. If we wish to see it straight we can use an ocular constituted by a concave lens, or insert a third one between the objective and the ocular (Fig. 5.18).

This is the principle underlying the optical microscope, which allows us to examine various details, even if they are minuscule (Figs. 5.19 and 5.20). How can we magnify an object? Are there limits? Certainly we cannot infinitely enlarge it, and, in particular, the optical microscope is unable to surpass a magnification factor of 400×. Why?

The human eye, without the aid of lenses, distinguishes up to 0.1 mm. If a detail of these dimensions is magnified 400 times, we see items of 0.1/400 = 0.00025 mm, or 0.25 μm (microns).

Details of this dimension are smaller than the wavelength of light with which we observe them, which varies from 0.4 μm (violet) to 0.7 μm (red). Under these conditions, due to its wave nature, light undergoes diffraction, and the sharpness of the forms is lost. When luminous rays pass through a hole of a size comparable to their wavelength, the light beam does not retain the shape of the opening but widens itself. This broadening of the luminous beam

Light Techniques 233

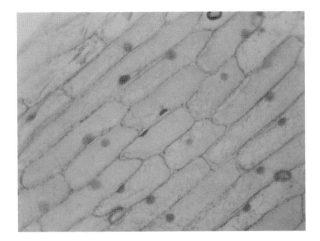

FIGURE 5.19 The peel of an onion magnified 100 times

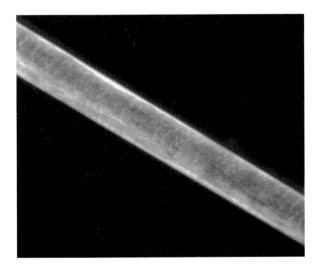

FIGURE 5.20 A human hair magnified 200 times

produces blurred images and is not only caused by holes but also by slits or steps, elements that constitute the details of a particular structure (Fig. 5.21). That's why the light microscope does not exceed 400 magnification.

However, it is possible to go beyond. Several solutions have been found that allow us to overcome this limit. Let's see some of them.

234 Visible and Invisible

FIGURE 5.21 Luminous rays striking the edge of the blade generate a light that widens due to the effect of diffraction

The wavelength of ultraviolet light is lower than that of visible light, and diffraction takes place with smaller dimensions. This allows for magnifications higher than 400 times. The images acquired with a microscope in the ultraviolet range are not directly visible to the human eye but can be observed through a fluorescent screen.

In the electromagnetic spectrum, after the ultraviolets the X-rays are found, characterized by wavelengths lower than 0.01 μm, and they are therefore able to achieve very high resolutions. Microscopes operating in the ultraviolet (or x-ray) range do not provide the light directly to the eye, but generate a signal that is processed by the supporting electronics, producing a visible form on a screen.

The question then naturally arises is why limit ourselves to investigating the micro world with electromagnetic radiation? Are there other probes capable of investigating the extremely small? Certainly. A microscope of widespread use is the electronic one, which uses a beam of electrons. Even these particles can produce diffraction, but their wavelength can be reduced to such an extent as to allow microscopic analysis with extremely high resolutions (Fig. 5.22).

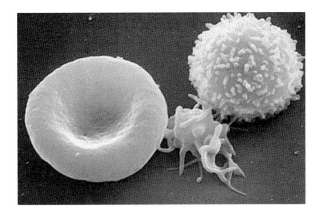

FIGURE 5.22 Some human blood components observed through an electron microscope. From *left* to *right*, we can make out a red blood cell, a platelet, and a white blood cell

Photometry

Light can be measured by taking various quantities into consideration. The power of a source, measured in watts, is obtained by dividing the energy that it transmits by its time of emission. Over an equal time, the most powerful lights produce the greatest illumination. By dividing power by the amplitude of the illuminated area, a new quantity is obtained, called irradiance, which is measured in watts per square meter.

Power and irradiance are detected by instruments that do not consider the sensitivity of the human eye, but refer to all electromagnetic waves, even those that are invisible. Being independent of our sight, these two quantities are classified as radiometric.

In everyday practice, the degree of illumination is often related to our sensations. A room is well lit when we clearly see the objects it contains. The human eye is a very special detector, giving greater value to certain wavelengths. Human sight is sensitive to a narrow portion of the electromagnetic spectrum—that of visible light—with maximum response to the yellow-green wavelengths. As a result, it is appropriate to measure light also on the basis of our reactivity, introducing a second category of indicators—photometric quantities (Fig. 5.23).

236 Visible and Invisible

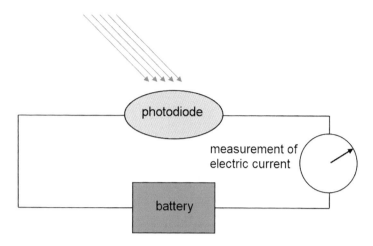

FIGURE 5.23 Representation of a light meter, a photometric instrument used to determine the amount of light present in an environment, particularly in photography and cinematography. A photodiode, i.e., an electronic component that conducts only when illuminated, is exposed to luminous rays and reacts, allowing for the passage of more or less electrical current in the circuit in which it is inserted. That current is then converted into a numerical value that appears on the instrument display

In photometry, the established indicators are luminous power and illuminance. The former is measured in lumens, the latter in lux, i.e., lumen/m². Both are defined only for visible light and respectively describe the power and the illumination level, taking the sensitivity of the human eye into account (Fig. 5.24).

The choice between radiometric and photometric quantities depends on the context. To install a photovoltaic panel, we examine the light that will shine on it in terms of radiometric quantities, while when designing an interior lighting system, it is appropriate to use photometric parameters.

The Temperature of Color

Different sources can spread visible light—the Sun, Moon, stars, lamps, incandescent bodies, etc. The light produced differs in terms of hue and, in order to specify them, an indicator that characterizes each source is used. Color temperature is measured in Kelvins (K). This unit of measurement is obtained by adding the constant 273 (273.15 for precise calculations) to degrees Celsius (°C).

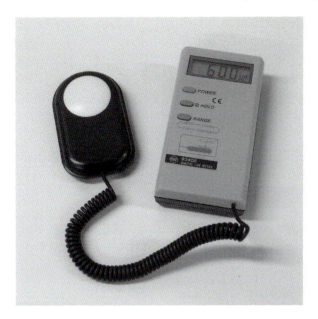

FIGURE 5.24 In traditional light meters, the sensing element was formed by a cadmium sulfide photo resistor, while now photodiodes based on silicon or gallium arsenide are used

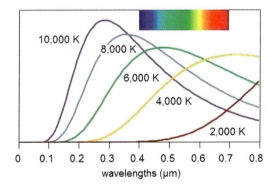

FIGURE 5.25 Variation in the emission of five different black bodies

Every light source is characterized by many features: temperature, constituent material, shape, surface, etc. On the other hand, for a black body, i.e., for any object that absorbs all incident radiation, the light emitted depends only on its temperature.

The graph in Fig. 5.25 shows the emission of five black bodies at 2,000, 4,000, 6,000, 8,000 and 10,000 K, respectively.

All the bodies considered diffuse in the range of visible light, shown in the figure by a colored strip. The hue depends on the relative weight of the different wavelengths. The emission is predominantly red if the source is located at 2,000 K, while it is white between 4,000 and 6,000 K, when all colors are diffused. In the high temperature spectra, violet and blue predominate.

It is possible to associate a well-defined tone to each black body temperature, thereby classifying the light sources (Table 5.1). For example, when the light produced by a lamp has the same hue as that generated by a black body at 3,000 K, it is said to have a color temperature of 3,000 K.

At lower temperatures we have the warm colors, while for values greater than 5,000 K we find the cold ones. Notice the oxymoron. 'Cold' lights correspond to higher temperatures (Fig. 5.26).

TABLE 5.1 Color temperatures of various sources

Source	Color temperature
Match flame	1,700 K
Candle flame	1,850 K
Incandescent lamp	2,700 to 3,300 K
Screen of a cathode ray tube	9,300 K
Moonlight	4,100 K
Light of a cloudy daytime sky	6,500 K
Light of the clear daytime sky	10,000 to 18,000 K

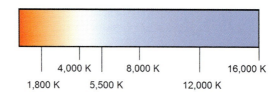

FIGURE 5.26 Scale of color temperatures

COLOR CODES

The various bright hues of fluorescent lamps are identified by a color code, made up of three digits. The first contains information on the ability of the light source to represent colors (9 for very good color rendering, 8 for good color rendering, etc.), while from the second and third one can determine the color temperature of the diffused light. In fact, the last number is obtained by adding two zeroes to these digits (Table 5.2). For example, 835 indicates a tri-phosphor fluorescent lamp (for the digit 8) with a color temperature of 3,500 K (for digits 3 and 5).

TABLE 5.2 Codes and Hues of modern lamps

Second and third digit of the color code	Color temperature	Characteristic of the light diffused
27	2,700 K	Extra warm light
30	3,000 K	Warm white
35	3,500 K	Neutral white
40	4,000 K	Cool white
65	6,500 K	Daylight
80	8,000 K	Sky white

Thermography

Thermography is a technique for acquiring images in the infrared range, generating a picture in which the color specifies the local temperature of the object analyzed. The instrument used to create a thermal image is the infrared camera that, after detecting radiation, converts it into visible light (Figs. 5.27 and 5.28).

Thermographic images are nothing but maps of infrared radiation emitted from the bodies, and to obtain them it is not necessary for the bodies to be illuminated. On this basis, it can be used as a nocturnal viewer to "see" in the dark, and this characteristic is also exploited by some predators. The body of a warm-blooded animal, in fact, has a temperature greater than that of the surrounding environment.

240 Visible and Invisible

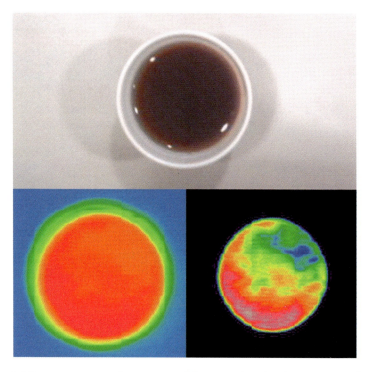

FIGURE 5.27 Let us consider a cup of hot tea (*top*). The images obtained with infrared imaging vary according to the interval of temperatures considered. On the *bottom left* the interval is wide; *blue* corresponds to 10 °C, *red* to 60 °C. The colors of the table, the cup, and the liquid correspond to approximately 20, 35 and 56 °C respectively. On *bottom right*, the interval is narrower; table and cup are out of range (too cold) while the hot liquid has a temperature varying between 54 °C (*blue*) and 58 °C (*red*). The colors detected by the infrared camera continuously rotates, because the different parts of the hot liquid move in a turbulent manner. This motion is an effective way to disperse heat and thus to reach thermal equilibrium with the environment

Thermal equilibrium implies a process of continuous emission and absorption of radiation that, at usual temperatures, takes place in the infrared region. This is the basis on which thermography is based. If a body is heated, this exchange may also take place in the visible range, with the emission of directly observable rays.

When iron melts, the peak of maximum diffusion is still found in the infrared range, even though the emission curve tails into the visible range. This is why incandescence can be seen at around 1,000 °C.

FIGURE 5.28 Let us now examine the image of a person, in this case the author of this book, seen through visible light and infrared rays. The *dark blue* corresponds to 17 °C, the *yellow* of the sweater to 28 °C, while the colors of the face are located in an interval between 33 and 37 °C. On the *top*, instead, the colors are distributed across a narrow range of temperatures, which highlights the details

CONSIDER THIS

In thermography, we transform infrared rays into visible ones, i.e., we modify their frequency. This however is not the only case in which electromagnetic waves undergo variations. As a result of fluorescence or phosphorescence, ultraviolet rays can become violet, while blue may appear green. Why is a suitable material sufficient in these cases, while in thermography a complex machine is required, connected to a battery?

Diagnostic Imaging

Our body is an extraordinary system consisting of many organs that we need to preserve, take care of, and treat when necessary. Only recently have we succeeded in an enterprise that in the past was considered impossible—to see our internal parts.

This possibility has greatly enhanced diagnostics, namely the medical survey that is carried out to identify possible diseases or dysfunctions. In modern medicine, imaging diagnostics is a crucial frontier, one in continuous development.

How is it possible to see inside our body, which is opaque and therefore does not allow for the direct observation of its constituents?

Some internal regions can be observed with a technique called endoscopy, which uses optical fibers to illuminate and convey images to an external display through a micro camera inserted inside the body. This system is the basis of the optical instrument used in video surgery operations.

Is it possible to go further and to enable the transparency of the human body, thus seeing the internal organs directly? Of course, as long as we are not limited to visible light.

Electromagnetic radiation features a range of characteristics, as its wavelength varies. The transparency of a body is not an absolute property but depends on the type of radiation considered. For example, the walls of a house are opaque to visible light but transparent to radio waves.

For over a century it has been known that X-rays pass through the human body, and radiography, which uses such rays, has become an important diagnostic technique, developed alongside computerized axial tomography, or CAT, in which the X-ray emitter rotates around the patient (Figs. 5.29 and 5.30). In this way, we obtain a sequence of data that, through a computer, allows for a three-dimensional representation of the organs analyzed, which may thus be rotated or sectioned at will (Fig. 5.31).

Light Techniques 243

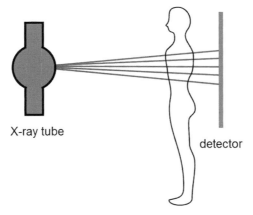

FIGURE 5.29 During radiography, X-rays are used that, by crossing the patient's body, are absorbed in different ways by the various tissues, then recorded onto photographic film (detector). Currently so-called digital radiography it also used, a term that means using the techniques of digital image acquisition, such as software and hardware deployed for the storage and processing of the images acquired

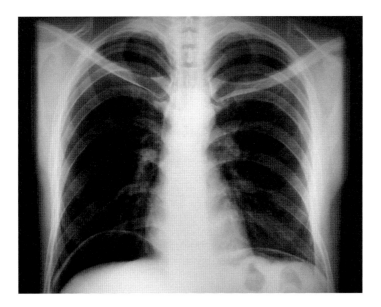

FIGURE 5.30 Biological tissues are crossed by X-rays with a partial absorption, depending on the material encountered by the radiation. For example, soft tissue transmits them more than osseous ones

244 Visible and Invisible

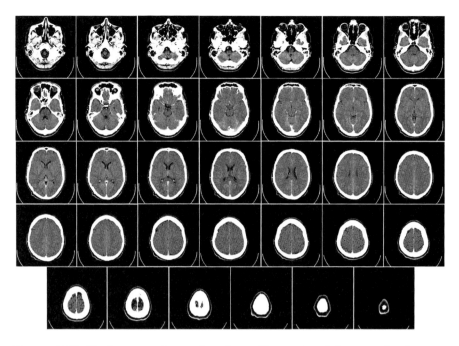

FIGURE 5.31 Various sections of a three dimensional image of a human skull from the base to the top of the head, produced using a CAT

When we have a chest X-ray, we absorb a certain amount of radiation, equivalent to that emitted by the natural environment in about a month. If we undergo a CAT scan, since it is a sequence of radiographic images, we absorb a higher dose. A computerized axial tomography of the skull exposes us to the same amount of radiation emitted from the natural environment in 25 months. Examinations of this kind, therefore, should be carried out only if they are strictly necessary.

Ultrasounds have higher frequency and are therefore not audible to the human ear. Ultrasonography uses ultrasound with frequencies between 2 million and 20 million hertz. Acoustic waves are directed towards the organ to be examined, from which they are partially reflected. In such a way, it forms an echo, delayed compared to the signal, making the two separable. The reflected wave contains information about the bodies that generated it and thereby allows for the reconstruction of the shape of the internal organs. This methodology does not use ionizing radiation and

> **ULTRASONOGRAPHY AND NUCLEAR MAGNETIC RESONANCE**
> Electromagnetic waves are not the only ones that provide us with a representation of the internal organs of the human body. Other techniques have been developed, each with specific characteristics and with a well-defined field of action. Let's look at two of them. Sound is a mechanical wave that propagates with the vibration of the matter in which it travels; like light, sound waves vary in space and time. Our ear is sensitive to sounds ranging from 20 to 20,000 Hz.

has no side effects. Therefore, its use is very widespread, although it is not able to render all anatomical details.

Another very powerful technique is nuclear magnetic resonance, capable of identifying the atomic species present in tissues. With this analysis, it is therefore possible to reconstruct more detailed images. To acquire a map, the patient is immersed in an intense magnetic field, which partially orients the atomic nuclei of the part examined. Radio waves are then sent that are absorbed differently by the various atoms, allowing their distribution to be represented. This examination does not use ionizing radiation either, but care should still be taken due to the intense magnetic field used, incompatible with the presence of metallic objects.

Surgery with Images

Having enabled us to observe the interior of the human body with an instrument, medical technology has gone one step further, adding devices useful for surgical procedures.

Obviously, this comes in the form of tiny tools (pliers, scissors, etc.) that the surgeon can manipulate from the outside, which are introduced through holes of 1 to 2 cm (Fig. 5.32).

In this case, the surgical operation is not observed directly, as in open surgery, but viewed through the optical image transmitted

246   Visible and Invisible

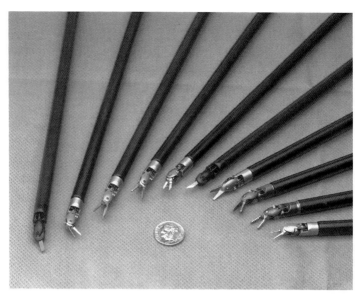

FIGURE 5.32 The instruments used for video surgery have extremely small dimensions

from the instrument. For this reason we speak of video surgery. The main advantage of this technique is its minimal invasiveness, thanks to surgical access using three or four small incisions. Video surgery therefore stands out by virtue of its mild postoperative pain, rapid recovery, and shorter hospital stay.

However, to perform an intervention using video surgery, considerable skill is needed. One of the main difficulties is the absence of three-dimensionality that the video image entails. It is difficult to operate without a clear grasp of distances.

Our perception of depth arises from the fact that the two eyes see shapes slightly staggered, and it can be recovered if the surgeon is provided with two independent figures, using the technique known as stereoscopy.

VIDEO SURGERY AND STEREOSCOPY
Video surgery instrumentation (Fig. 5.33a) has recently been developed that exploits the sensation of depth through an optical instrument which, in addition to lighting, carries two independent images: one for the right eye and one

for the left, as shown in Fig. 5.33b. The screen is made up of a stereoscopic viewer, which allows for complete three-dimensional observation (Fig. 5.33c). In this way, the surgeon can operate with the same visual perception that may be provided by direct observation.

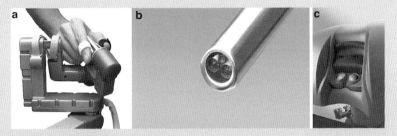

FIGURE 5.33 During an intervention of video surgery the surgeon maneuvers the instruments from the outside, through sophisticated devices, such as those shown here

LEDs

The traditional incandescent lamp is highly inefficient, since 98 % of the energy it consumes is diffused in the form of heat. For this reason, many studies have been undertaken to develop more efficient lamps.

A promising source is offered by the LEDs, an acronym for light emitting diode. These devices are made up of semiconductor materials, similar to those used to manufacture electronic devices that we use daily, from PCs to mobile phones.

The LED does not bring a filament to incandescence, as in the traditional lamp, nor does it trigger an electrical discharge, as in fluorescent light. Instead it exploits the properties of electrons in different layers of the device. These absorb energy from the electric power supply circuit, to return it almost entirely in the form of electromagnetic radiation. It is a very efficient emission that produces cold light.

From many points of view, LEDs are superior to all other types of lamps, not only to incandescent ones. For example, compared to fluorescent lights, LEDs light up instantly, their lifetime

248 Visible and Invisible

FIGURE 5.34 An example of LEDs used in traffic lights

is not affected by on-off cycles, and they do not contain mercury and are less fragile, since they use solid-state devices, not glass tubes (Fig. 5.34).

Is it possible to quantify the superiority of LEDs? To answer this question, we need to measure the efficiency of various lamps.

Any light source may be regarded as an apparatus with an input, where electric power enters, necessary for its functioning, and an output, from which the luminous power emitted comes out (Fig. 5.35). The lamps considered to have a good yield have a high luminous output while consuming little electric power. The efficiency of a lamp is thus given by the ratio between the first, measured in lumens, and the second, given in watts. We may note that photometric quantities are used, because the illumination of an environment depends on the visual sensitivity of the human eye. Consider, for example, a 60 W incandescent lamp, which emits about 870 lumens. Its efficiency may be calculated as $870/60 = 14.5$ lumens/Watt.

Fluorescent lights can reach values much higher, from 46 to 75 lumens/Watt. For this reason they have largely replaced incandescent bulbs. Various LEDs, on the other hand, have yields ranging from 55 to 93 lumens/Watt. The upper value corresponds

FIGURE 5.35 The efficiency of a lamp, like any light source, is given by the ratio between the luminous power that it emits and the electrical power it requires

FIGURE 5.36 A luminous top by the French firm LumiGram

to a model with the E26 screw base, which absorbs 8.7 Watt and produces 810 lumens. Research in this field is progressing fast.

LED technology is, however, not only profitable in the field of traditional lighting, but it appears more and more promising in generating new products and ideas (Figs. 5.36 and 5.37).

The peculiarity of LEDs is their ability to transform electrical energy into luminous rays with wavelengths within a narrow range. These lamps do not emit white light but light of a single color, determined by the constituent material. The first emission was achieved in the infrared range with LEDs based on gallium arsenide. The red light is obtained by using gallium arsenide and aluminum, or by inserting atoms of zinc in gallium phosphide.

250 Visible and Invisible

FIGURE 5.37 The British company LOMOX announced the sale of luminous walls based on organic LEDs

If the same compound is enhanced with nitrogen atoms, green emission is obtained. For blue, either silicon carbide or an artificial crystal, called a quantum well, are used, in which ultrathin layers of two materials alternate—nitrides of gallium and indium.

By combining red, green, and blue, any colored light may be obtained. There are manufactured LEDs that emit ultraviolet light. Furthermore, if a thin layer of phosphorus is deposited on a device that generates blue light, part of the light produced changes color due to phosphorescence. Acting on these layers of phosphorus, white and several other colors have been obtained.

THE FIRST LED CITY

Torraca, an Italian town in the Cilento area, is the first municipality in the world to have installed a public lighting system using exclusively high-efficiency LED technology (80 lumen/Watt). These lamps were placed in more than 700 points around the town, along all municipal access roads and in the characteristic alleys of the old town. The *Economist* magazine named Torraca the first 'LED city' in the world.

Photovoltaic Panels

The most usable form of energy is electric power. It can produce light, sound, motion, heat, images, and is the only type of energy suited to use in computers and communication technologies.

Light can be converted into electric energy directly with a photovoltaic system. In this device, electrons convert power from luminous to electric form, with a process that is similar but opposite to what takes place in an LED (Fig. 5.38).

By illuminating an LED, electricity is produced. Obviously, the materials and design of modern photovoltaic devices are different from those of LEDs, being optimized for the production of electric power.

The photovoltaic panel is a collection of many units, known as solar cells. Each element, such as that shown in Fig. 5.39, is covered by a metal grid which carries the electrical charge produced, and each individual cell is connected to the other by means of metal contacts.

The efficiency of a power plant is defined, like that of the LEDs and other light sources, as the ratio between the output/input power. In this case, however, the former is electric and the latter luminous. Both are measured in watts; in this case, in fact, radiometric quantities are used since the device is powered by sunlight.

Photovoltaic power is currently the energy production technology undergoing greatest development throughout the world.

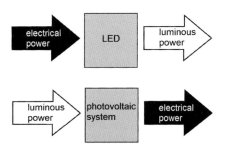

FIGURE 5.38 A photovoltaic system is nothing more than a LED traveling in the opposite direction, with luminous power as input and electrical power as output

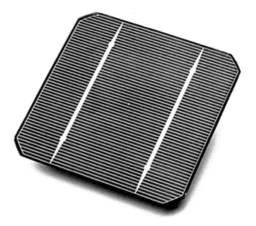

FIGURE 5.39 Solar cell from a photovoltaic system

> **THE FIGURES OF SOLAR ENERGY**
> How much of a role does the Sun's energy play in satisfying our needs? The annual consumption of humanity corresponds currently to about 15,000 billion watts. The luminous power that reaches Earth from the Sun is 6,000 times larger, amounting to some 90 million billion watts. So we have an abundance of solar energy, but we still need to be able to use it efficiently (Figs. 5.40 and 5.41).

FIGURE 5.40 A highly innovative architectural structure, the Solar Pergola in Barcelona, consists of nearly 2,700 photovoltaic modules covering an area equal to that of a football field

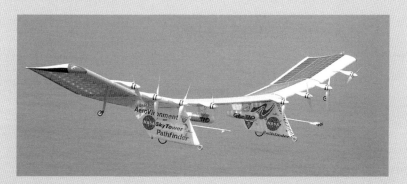

FIGURE 5.41 The NASA Pathfinder-Plus airplane, pilotless and powered by solar energy, in flight over Hawaii in June 2002

254 Visible and Invisible

TABLE 5.3 Better efficiency at lower costs

Generations	Features	Efficiency	Cost in $/Watt
First	Amorphous or crystalline silicon	12–14 %	4 to 6
Second	Thin films deposition	8–10 %	0.5 to 1
Third	Nano technologies	30–50 %	0.2

As can be seen from Table 5.3, much research is dedicated to the optimization of materials and the fabrication technologies of photovoltaic panels to achieve better efficiency at lower costs. We are now able to produce systems competitive with other energy sources, also from an economic point of view.

Optical Fibers

In the nineteenth century it was discovered that one could transport light, confining it within transparent substances such as water or glass.

Figure 5.42 illustrates a luminous fountain created by leaving some water to flow from a container. The lamp, placed behind the tank, illuminates the exit hole. When the room is darkened one realizes that the light is channeled into the jet of water, illuminating it. In this experiment the luminous rays follow the flow of the liquid by bending themselves. How is this possible? The explanation lies in the total reflection of light on the internal walls of the jet.

The confinement of light is due to refraction. When the light beam crosses the separation between the water and air it is deflected towards the surface that divides the two substances, as shown by the dashed path in Fig. 5.43. If the inclination of the beam is large (trajectories dotted and solid), the change of direction does not allow the rays to pass through the interface, and they continue in the water. In the case of a jet of water, comparable to a cylinder, the beam continues to bounce along the jet's internal surface.

Optic fiber technology is based on this effect. The light is injected into the fiber at an angle such as to ensure the confinement

Light Techniques 255

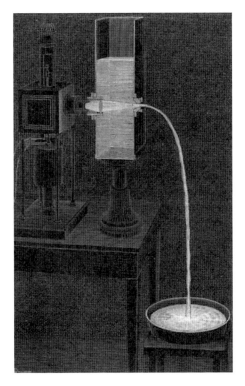

FIGURE 5.42 This fountain was created by letting water flow from a container; drawing by the physicist Jean-Daniel Colladon (1802–1893) and published in 1884 in 'The Nature' magazine

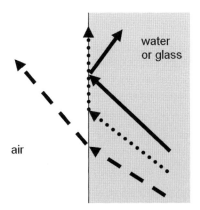

FIGURE 5.43 With an increasing angle of incidence, the refracted ray is deflected towards the surface of separation of the two media, until it disappears. In this case there is a total reflection of the incident ray

FIGURE 5.44 Optic fibers convey light through continuous reflections on their own internal walls. In this way they can carry optical signals over long distances

of the rays within filamentous structures of glass or plastic. These fibers are used in many pieces of equipment, including lamps, endoscopes, and optical instruments (Fig. 5.44).

The most significant application, however, are in the fields of information and communication technologies. Computer networks based on the optic fibers have the greatest capacity for data transmission, hundreds of thousands of times more powerful than microwave transmitters or satellites. In the future such fibers will be essential components of digital devices.

Transmitting Through Waves

The space where we live is never empty, even when it appears so to us. It is continuously crossed by radio signals, electromagnetic waves of long wavelength and low frequency. To verify this, simply turn on a radio, a television, a cell phone ,or a mobile modem. The signals we receive everywhere demonstrate the large number of these waves.

In the past it was not so. The terrestrial environment has always been crossed by radio waves, but much lower amounts than today. Their presence in nature is due to several causes, of which the main one is associated with the natural irradiation of every object. In the telecommunications sector, this effect is classified as background noise.

Most of the radio waves existing today are instead a consequence of the extraordinary development of telecommunications, which began at the start of the twentieth century. Finding a 'free' space, i.e., a frequency to transmit not yet used, is becoming increasingly difficult. Access is governed by laws, on the basis of defined bands based on their best use. For example, ELFs (extra low frequencies) and SLFs (super low frequencies) are the most suitable for transmission in salt water, while ULFs (ultra low frequencies) can penetrate the soil.

Guglielmo Marconi (1874–1937), on December 12, 1901, succeeded in sending a message from Poldhu in Cornwall (England) to St. John's, Newfoundland (Canada), using electromagnetic waves. The first transoceanic communication therefore overcame a distance of 3,400 km. In doing this, Marconi challenged the skepticism of colleagues of the time, who thought impossible the transmission of radio signals over long distances, believing that the curvature of Earth would have prevented the connection between the opposite shores of the Atlantic Ocean. The experiment went instead to a successful conclusion. The Italian scientist had unknowingly taken advantage of the properties of the ionosphere, then still a mystery, to reflect the radio signals and make them bounce to the other side of the ocean.

Radio waves are used to carry information, such as sounds and images. The simplest transmission technique is amplitude modulation (AM), the first system used in the telecommunications field. In practice this means modulating the amplitude of the radio signal that it uses to transmit (carrier) in proportion to that of the signal that it sends (modulation).

Now, imagine we want to send the musical note 'A-440,' which has an acoustic frequency of 440 Hz, with a radio transmission at 2,200,000 Hz. The 440 Hz acoustic signal that we want to send varies 5,000 times slower than the electromagnetic wave. Figure 5.45 shows two oscillations with different frequencies: high in the gray curve, low in the black one.

258   Visible and Invisible

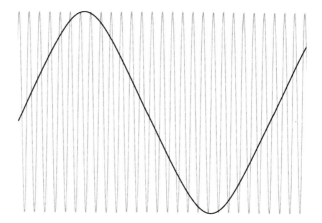

FIGURE 5.45 The different frequency of two oscillations: the signal to be transmitted in AM (*black*) and the electromagnetic wave that we need to use for transmission (*gray*)

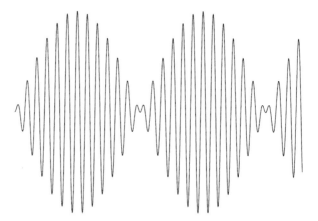

FIGURE 5.46 The modulated radio wave that will transmit the signal shown in *black* in Fig. 5.45

In amplitude modulation, the height of the radio wave is not constant but varies on the basis of the information carried. Figure 5.46 presents the modulated radio wave, which serves as a carrier of the audio signal.

In communications with frequency modulation (FM), the information does not change the amplitude of the carrier but its frequency.

> **CONSIDER THIS**
> A lighthouse marks a place by emitting visible light, while a radio beacon transmits a radio signal on a specific frequency. Why are both less used today than in the past?

Table 5.4 shows the range of radio waves divided into different bands, indicated by a code derived from the words that characterize them. As we saw earlier, ELF stands for extremely low frequency and the other acronyms are obtained in a similar way, according to the following terms: S (super), U (ultra), V (very), M (medium), and H (high). Wavelengths are expressed in meters and multiples and submultiples of the meter. Frequencies are specified in Hertz and its multiples. The terms AM and FM, refer respectively to 'amplitude modulation' and 'frequency modulation.'

TABLE 5.4 Radio bands, the section of the electromagnetic spectrum reserved for telecommunications

Band	Frequency	Wavelength	Some uses
ELF	3–30 Hz	100,000 to 10,000 km	Transmission with submerged submarines
SLF	30–300 Hz	10,000 to 1,000 km	Transmission with submerged submarines
ULF	300–3,000 Hz	1,000 to 100 km	Communication in mines
VLF	3–30 kHz	100 to 10 km	Transmission with surfaced submarines
LF	30–300 kHz	10 to 1 km	AM radio (long wave); aeronautical and maritime radio beacons;
MF	300–3,000 kHz	1 km to 100 m	AM radio (medium wave); avalanche transceivers
HF	3–30 MHz	100 to 10 m	Radio-amateurs; cell Broadcast
VHF	30–300 MHz	10 to 1 m	FM radio; television; aeronautical, civil, naval and police communications
UHF	300–3,000 MHz	1 m to 100 mm	Television; mobile phones; GPS; WLAN
SHF	3–30 GHz	100 to 10 mm	Radar, satellites, WLAN
EHF	30–300 GHz	10 to 1 mm	Satellite and amateur-radio transmissions

Radar

The eye, the camera, the video camera, and other optical instruments identify the objects only when they are lit up. The presence of light sources is needed to see them.

Radar instead detects images even in the dark, because it carries out both functions necessary to vision—it 'illuminates' the object and collects the waves diffused by it. To do so, it does not use visible light, which obstacles such as clouds or mist can absorb, but microwaves and radio waves. The weak return signal is then amplified by the receiving antenna and supporting electronics systems (Figs. 5.47 and 5.48).

With this technique it is possible to detect objects and determine their distance. The returning pulse, in fact, arrives later than that emitted, and the amount depending on the distance of the object. For each kilometer, the received signal is delayed by about 6.67 microseconds, a calculation based on the fact that electromagnetic waves travel at the speed of light.

Radar technology is often more effective than similar systems—i.e., sonar and sodar (Wind Profiler) radiometer—especially in the case of metal detection, sea water or wet earth. In addition, the

FIGURE 5.47 A typical radar instrument

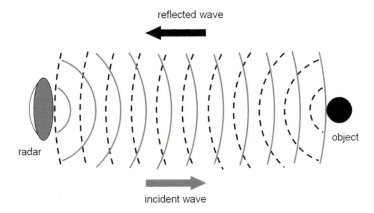

FIGURE 5.48 The radar sends a short pulse (*solid lines*). When the waves hit an object, they are partially diffused and a weak signal returns (*dashed lines*)

FIGURE 5.49 Image of Hurricane Isabel (September 18, 2003), mapped on the basis of the data gathered by the NWS NEXRAD radar in Morehead City, North Carolina

system can operate even in the presence of fog, clouds, rain, or snow. It is sufficient to remove the frequencies that are absorbed by water vapor, rain, or atmospheric gases from the radar signal.

However, in meteorology, to detect similar atmospheric agents, the opposite choice is made, using their absorption frequencies.

Radar is employed in civil aviation, in meteorological services, in remote sensing, in astronomical research, and in the military (Fig. 5.49).

Satellite Positioning

Today we can travel guided by a satellite navigation system that offers us an aerial view of the route, showing us the way to reach a given destination. Obviously, the navigation device must have detailed maps of the area, but this is not enough. It must also constantly monitor our route and know our position at any time to within a few meters.

All this is possible thanks to a satellite positioning system, whose operation is based on the extraordinary achievements of science and technology, including:

- The development of a network of spacecraft that transmit radio signals in the UHF band.
- An understanding of the properties of radio waves.
- The ability to engineer high-precision atomic clocks (with a margin of error less than 0.000002 ms/day, 0.7 ms in a thousand years), very sophisticated and expensive, to be installed on satellites.
- The possibility of continuously synchronizing the quartz clock of the receiver with the atomic clocks mounted on the satellites, thanks to the network of the navigation system.

Each spacecraft sends a regulation signal, which takes a certain amount of time to reach its destination. The clock of the receiver is able to maintain perfect synchronization because it is reached by data from at least four satellites. A single value of time, the right one, must in fact agree with the four different estimates.

Let us now examine the functioning of satellite navigation: the spacecraft transmits its position and the precise time of sending, and the traveler receives this information with a certain delay.

If the satellite is positioned at 20,200 km, the electromagnetic wave will take 67 ms to travel this distance. This looks like a difficult time to measure, but the accuracy of the receiver clock, continuously synchronized, is such to allow for its determination.

By knowing the delay it is easy to calculate the distance traveled by the radio wave; therefore the receiver knows the position

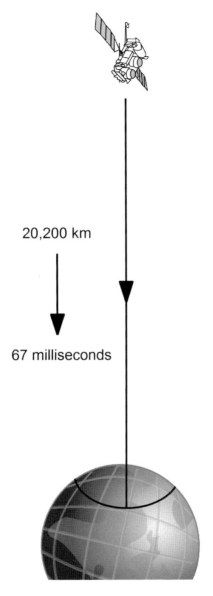

FIGURE 5.50 All the terrestrial locations situated on the *black line* are at the same distance from the satellite and receive its signal with the same delay

of the satellite and its distance. This, however, is not enough to know where the traveler is located, since all the places that are on the dark curvedline in Fig. 5.50 have the same distance from the satellite.

264   Visible and Invisible

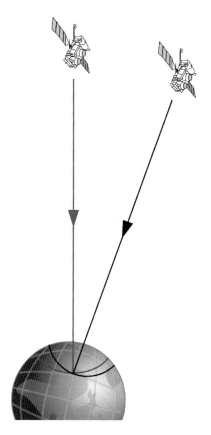

FIGURE 5.51 To properly locate the position of a traveler, more satellites moving along different orbits are needed

To dissolve the ambiguity it is necessary to repeat the procedure with other satellites, positioned on different orbits (the other curve in (Fig. 5.51). This is possible thanks to the navigation satellite system network. The spacecraft can send signals that allow the receiver to identify different curves. From their intersection the correct positioning is achieved. Four satellites are needed to determine the position of the traveler unambiguously.

The satellite navigation system most widely used is the one developed in the United States, GPS, the only one currently fully operational. The European Union has designed a system, called Galileo, which should become operational in 2015.

> **CONSIDER THIS**
> The clocks on satellites move faster in reduced gravity, because of their altitude. According to the theory of relativity, their time passes faster than on Earth, and advances each day by approximately 0.038 ms, thus soon losing synchronization between the two types of clocks. For this reason the electronic system has to apply the appropriate relativistic corrections. Is that true?

Fluorescence and Phosphorescence

If you enter a dark room, 'illuminated' by ultraviolet light, a new world of color awaits you. Your clothes and objects take on different and unusual tints (Figs. 5.52 and 5.53). The cause of these strange effects resides in the invisible ultraviolet rays present in the room, so even in a dark environment, some objects shine and therefore emit visible light.

Normally, a body diffuses a light equivalent to that which illuminates it; in this case, however, ultraviolet waves induce an emission in the visible range due to particular substances capable of absorbing a type of radiation and then diffusing a modified version of it, with a lower frequency. How is it possible for matter to transform light?

Electromagnetic waves are constituted by particles called photons, the energy of which is related to frequency. If the former decrease, so does the latter. The substances that convert ultraviolet light into visible light lower the energy of the photons. Part of the latter is dispersed in the material and the remainder is emitted as light of a lower frequency.

Nothing strange—energy is conserved. The presence of a substance capable of modifying light by increasing the frequency would instead not be understandable. There are no bodies that emit more energetic radiation than they receive.

266   Visible and Invisible

Figure 5.52 In an environment illuminated with Wood's lamps, which emit mainly ultraviolet light, shirts are lit due to the effect of the fluorine they contain

Figure 5.53 Bodies and materials light up, diffusing a range of colors

There are two types of material capable of transforming light: fluorescent and phosphorescent. The former describes those substances that cease to give off light as soon as the light that triggers them is turned off.

Phosphorescent material, on the other hand, continues to emit light even in the absence of the other light source, for a period of time ranging from a few fractions of a millisecond to a few hours, depending on the type of material (Fig. 5.54). For this reason, phosphorescent substances are often used in road signs and in safety instructions.

> **CONSIDER THIS**
> Modern light bulbs, which have replaced the old incandescent bulbs, contain an efficient source of ultraviolet light. The glass that covers them is coated with fluorescent material that transforms diffused light into visible light. Can that be true?

> **SEEING THE MICRO WORLD WITH FLUORESCENCE**
> Fluorescence (Figs. 5.55 and 5.56) allows us to increase the capacity of microscopes for the investigation of the biological world. Indeed, it is possible to render fluorescent the various components of a structure, each with a different color, thus obtaining clearly highlighted images (Fig. 5.57).

268 Visible and Invisible

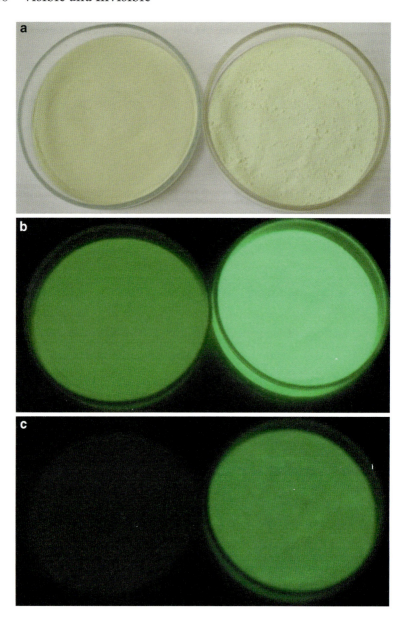

FIGURE 5.54 Two different phosphorescent pigments illuminated by visible light (*top*) are then placed in a dark environment (*middle*). The *bottom* picture shows them after 4 min. The light emission persists for both, but with different effectiveness

FIGURE 5.55 Many natural stones are fluorescent. The image here shows crystals of autunite under ultraviolet light

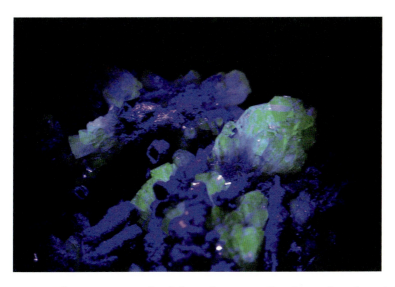

FIGURE 5.56 These are crystals of the substance adamite under ultraviolet light

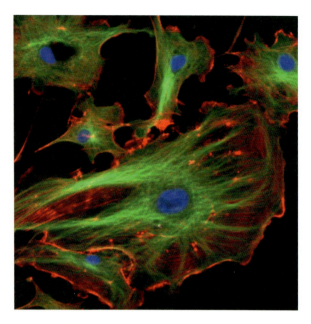

FIGURE 5.57 Cells of the inner lining of arteries and veins (the endothelium) can appear under a fluorescence microscope. The nucleus emits *blue* light, while the microtubules and the actin filaments diffuse *green* and *red* light, respectively

Lasers

Science and technology have found a way to generate a very special light, the laser, with particular features that make it unique. Its rays are highly collimated, which means they travel parallel to each other and the beam does not disperse. A laser is very bright and is able to carry a lot of energy; the beam is monochromatic, with only one frequency. The waves travel in unison or, in other words, are coherent.

The light of a normal lamp is not monochromatic but contains different colors, thus seeming white. Moreover, it is not coherent. The different waves are out of phase with one another, as they originate from atoms that emit independently (Fig. 5.58).

FIGURE 5.58 In a normal lamp the emitted light is not monochromatic nor coherent

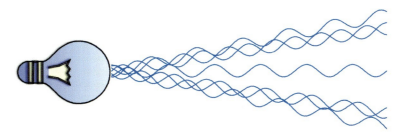

FIGURE 5.59 Coloring the glass or using discharge lamps we can obtain a monochromatic, but incoherent beam. The different waves do not reinforce each other

FIGURE 5.60 Thanks to the phenomenon of stimulated emission the laser light is coherent and monochromatic

If the light beam is colored, we can achieve monochromaticity but not coherence. The various waves do not reinforce each other (Fig. 5.59).

Laser light, as we have seen, is monochromatic and coherent, the different waves being emitted in phase, capable of adding up in the most efficient ways. Diffusion takes place in a cavity where radiation resonates with the atoms emitted, resulting in stimulated emission, i.e., caused by pre-existing radiation. This phenomenon was discovered by Einstein (Fig. 5.60).

272 Visible and Invisible

FIGURE 5.61 The laser beam sent to the Moon by the McDonald Observatory, University of Texas

The unique properties of the laser beam are at the foundation of many applications in diverse fields, from telecommunications to medicine, welding to measuring instruments.

About 40 years ago, the astronauts of *Apollo 11* deposited a set of mirrors on the Moon. This apparatus then allowed astronomers to send a laser beam from Earth, hit the reflective system, and receive the reflected signal. The laser light took about 2.5 s to complete the round trip. In this way it was possible to determine the Earth-Moon distance with extreme accuracy (with a margin of error of less than 3 cm). This value varies slightly during Earth's rotation, with an average of 385,000 km. In this way it has also been discovered that the Moon is moving away from Earth at a speed of 38 mm per year (Fig. 5.61).

> **CONSIDER THIS**
> The light saber, the energy of which is made up of light, is the most fascinating weapon in the *Star Wars* film saga. In the future, could these devices become real? For the time being, we shall content ourselves with examining some issues about it. The laser beam comes to an end in a vacuum, i.e., it's only present for the length of the sword. Could that happen? We know light rays are reflected by a metal surface. If an opponent wields a reflecting shield, would the light saber reflect back and hit its wielder? Light beams are known to intersect without problems. How might the light sabers collide with each other? Finally, a laser beam is invisible; only the objects illuminated by it can be seen. How would it be possible to see the twists and turns of the light saber in a vacuum?

> **THE SUPER LASER**
> It is possible to create a very powerful light through the use of the free electron laser, a device that overcomes many limitations of normal lasers. Its wavelength is not fixed, but it can vary considerably, from microwaves to X-rays. For this reason, and for other specific properties, such a laser provides a superior light, like the one created in the most advanced synchrotrons.

Synchrotron Radiation

There exists a light even more powerful and versatile than the laser. It is generated by a special machine, the synchrotron, and for this reason that light is called synchrotron radiation, and it allows scientists to carry out unique investigations in the micro (one millionth of a meter) and nano (one billionth of a meter) worlds.

FIGURE 5.62 Aerial view of the ESRF, an international research center supported by 19 countries. It has an annual budget of approximately 98 million euros, employs more than 600 people, and every year it hosts more than 7,000 researchers

At present we are witnessing the growing spread of synchrotrons, although they are very expensive to build and maintain. According to lightsources.org, around the world in 2011 there were forty major synchrotron facilities, and their distribution indicates how much interest in this equipment is universal: thirteen in Europe (four in Germany, two in France, one each in Italy, Russia, England, Denmark, Sweden, Spain, and Switzerland); ten in the Americas (eight in the USA, one each in Canada and Brazil); fifteen in Asia (six in Japan, three in China, one each in Russia, India, Korea, Thailand, Taiwan and Singapore); and one in Jordan and one in Australia.

The largest installations are at the ESRF (European Synchrotron Radiation Facility in Grenoble, France), at the APS (Advanced Photon Source, near Chicago in the United States), and the SPring-8 (Super Photon Ring-8GeV in Japan). Italy is equipped with a modern and competitive structure, called Electra, based in Trieste.

The heart of any synchrotron is a ring in which electrons travel at great speed. Being equipped with mass, these particles cannot reach the speed of light, although they come very close to it, up to 99.9999985 %. To produce intense light the ring must be large in size. The length of the one in Grenoble is 844 m (Fig. 5.62).

Where does the light come from? The radiation is created by 'bunches' of electrons in motion inside the synchrotron ring. Their orbit is not circular but formed by a series of straight trajectories.

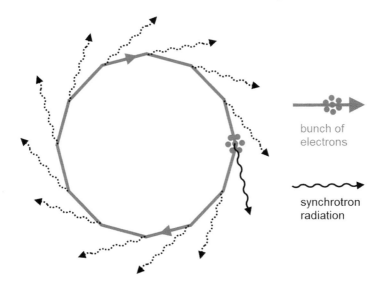

FIGURE 5.63 The orbit of a 'bunch' of electrons, formed by 12 straight trajectories (shown in *gray*). At every deviation the particles emit electromagnetic waves, i.e., synchrotron radiation (shown in *black*)

At the end of each of these, the particle beam is deflected by a series of magnets, and for this reason it emits electromagnetic waves. The electric charges diffuse synchrotron radiation at each curve, almost always in the form of X-rays. The luminous flux is then collected by a series of devices and used in the experiments. In Grenoble, forty beam lines are operative (Fig. 5.63).

The X-rays produced by a synchrotron may be 100 million times brighter than those created by a conventional source (Fig. 5.64).

Of what use is this light? It can help us to investigate the micro and the nano-world with unprecedented efficiency and versatility. Thousands of valuable research projects are developed every year with this instrument in mind (Fig. 5.65). Let us look at some results, for the sake of example only.

Synchrotron radiation is fundamental to the study of the still unknown properties of materials, from semiconductors to magnets, superconductors to nanostructures. It has been possible to obtain images of gold particles, 100 nm in size, offering details of 5 nm, just 30 to 40 times the size of an atom of this element.

This radiation not only enables us to analyze atomic processes but also to study their evolution over time. For example, it has

FIGURE 5.64 Synchrotron radiation that comes out from a beam line. The radiation is so intense it ionizes air molecules, which color the path of the rays

allowed us to investigate the rearrangement of a molecular bond after its breakage, despite the fact that this happens over a time span measurable in tenths of a picosecond. This phenomenon occurs with an unfathomable speed; 1 ps is in fact equal to 1 millionth of a millionth of a second.

The use of synchrotron radiation has made it possible to determine the protein structure that allows viruses to invade a host cell and merge with it, accurately showing how the infectious agent transforms to activate contagion. It has helped us to understand how the human immunodeficiency virus (HIV) establishes itself in the human body, by determining the structure of the enzyme used to copy the genetic information in the cell infected with the viral agent. The molecular configuration of a crucial part of the cellular receptor NMDA has been identified, which is involved in severe neurological diseases such as Alzheimer's and Parkinson's. The mechanisms of DNA repair in human cells have been studied, which is highly significant given that this molecule undergoes several changes on a daily basis. Using synchrotrons has allowed us to discover how to identify the presence of cells affected by a very aggressive and malignant form of brain tumor. The molecular mechanisms capable of fighting the parasite responsible for African sleeping sickness, which kills about 30,000 people each year, have

FIGURE 5.65 Several paintings of the past appear to us as altered from the original version because of the deterioration of their colors, due to several causes that scientists are only now beginning to understand. In particular, the radiation of the Grenoble synchrotron has been used to investigate the degradation of the *yellow* color that threatens some famous masterpieces by Vincent Van Gogh. A group of scientists from four countries, analyzing two works by the artist, "The Banks of the Seine," shown in the figure, and "View of Arles with Irises," identified the chemical reaction (and the conditions that foster it) underlying the transformation of *yellow* chrome, so dear to Van Gogh, into *dark brown*. This discovery is a first step towards the development of restoration techniques able to safeguard the most famous paintings by Van Gogh and many other Impressionists such as Seurat, Pissarro, Manet, and Renoir. In this research group, a central role was played by an Italian student, Letizia Monico

been precisely defined using these devices, and the microscopic processes that enable vegetable organisms to resist environmental stress such as drought and cold have been identified. We now know how plants change the distribution of arsenic in contaminated alluvial soil by accumulating it around their own roots because of this technology. Using synchrotrons the chemistry of the feathers and bones of Archaeopteryx, a creature halfway between bird and dinosaur that lived 150 million years ago, has been studied, as well as the fossilized teeth of children from the Neanderthal species, with the result that we now know that their adult teeth grew in earlier than those of Homo sapiens. Our prolonged childhood may have provided us with a reproductive advantage.

In the industrial field, synchrotron radiation is used for the manufacture of increasingly compact devices; among other things, it has led to the development of new materials for photovoltaic cells and the discovery of new processes for the production of photovoltaic energy.

The results obtained in biology are countless. With synchrotron radiation it is possible to perform X-ray radiographs and very detailed CT scans, producing three-dimensional images of biological cells without the distortions caused by lenses and at resolutions so high as to be able to examine the internal details.

The research on elementary particles makes use of large plants, among which the LHC stands out—the Large Hadron Collider at CERN (Conseil Européen pour la Recherche Nucléaire, European Organization for Nuclear Research). The LHC is a synchrotron designed to accelerate ions and protons up to 99.9999991 % of the speed of light and then make them collide. In this way it is possible to explore the extremely small, with particles one billionth of a nanometer in size and locally recreate the conditions found just after the Big Bang.

To achieve this goal, the largest scientific instrument in the world was built, with a 27-km-long underground circular ring, the inside of which is brought to an extremely high vacuum with a residual pressure ten times lower than that on the Moon. Superconducting magnets are placed along the tunnel containing the ring, cooled to –271 °C, a temperature lower than that of the cosmic vacuum (–270 °C).

The LHC also emits synchrotron radiation, but to a much lesser extent compared to facilities designed specifically for this purpose. The CERN machine does not accelerate electrons but heavier particles, and, consequently, the radiation emitted is lower. If the LHC protons were replaced by lighter electrons, the emission of electromagnetic energy would be 10,000 billion times greater. However, since its purpose is to accelerate the beams of particles, this loss of energy in the form of light should be avoided. Conversely, systems that produce synchrotron radiation make use of electrons, since their objective is to maximize these emissions (Fig. 5.65).

Photonics

In the second half of the twentieth century, a major change came about, the effects of which would profoundly affect our lives—the electronic revolution and the ensuing digitalization of our world.

In the digital world, information is stored and processed in binary form, as a sequence of 0 s and 1 s, i.e., off and on. Electronic devices are based on a large number of switches that can be turned on or off, called transistors. These are the real protagonists of digital revolution, and it is estimated that every year around 6 quintillion are built, nearly one billion for each inhabitant of Earth.

The first transistors were made with vacuum tubes, such as that used by Röntgen to study cathode rays. The breakthrough started when it became possible to produce them with solid materials, using techniques capable of making ever smaller systems. At present, the basic material used to manufacture electronic circuits is silicon.

The first integrated circuit, built in 1958, was made up of ten elementary components. Since then the research has led to an incredible miniaturization, well described by Moore's law: "The number of transistors that can be placed on an integrated circuit doubles approximately every 2 years" (Fig. 5.66).

For some time, experts have been aware of the existence of a limit to this growth. Miniaturization cannot continue indefinitely. The main problem is the production of heat by the electric current flowing through the devices. As electrons move from one transistor to another through connections, the number of transistors increases, and so does the length of the interconnections in the circuit.

What does it mean to have 10 km of electric cables miniaturized in a tiny plate? Since the electric current is carried by electrons that interact with the lattice of the solid, releasing heat, the chip acts like a piece of iron—it warms up. This is a well-known phenomenon for those who use a PC.

The electrical current flowing through the devices may be minimized, but the heat generated increases due to the progressive increase of the length of interconnections. We have almost reached the limits of the miniaturization process. Without any

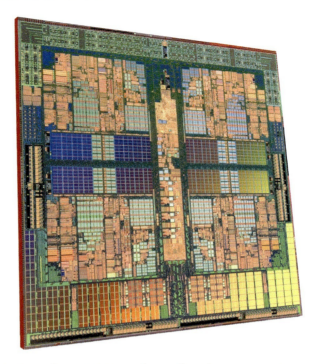

FIGURE 5.66 A modern AMD product, the Phenom quad-core microprocessor that, mounted on a pltate 15 mm long and 17 mm wide, contains 758 million transistors

intervention, by 2016, we would reach the point at which microprocessors would give off a level of heat equal to that emitted by a portion of the Sun's surface of the same size!

There is therefore an interconnected bottleneck that prevents the development of electronics. The various expedients identified to mitigate this problem are helpful, but do not remove it. Only a radical change may allow for further progress, a shift from electronics to photonics, which makes use of photons instead of electrons. With the use of light, the interconnected bottleneck is removed, since luminous rays flow through optical fibers without losing heat, and electrical connections are replaced by conduits in which electromagnetic waves travel at the speed

Light Techniques 281

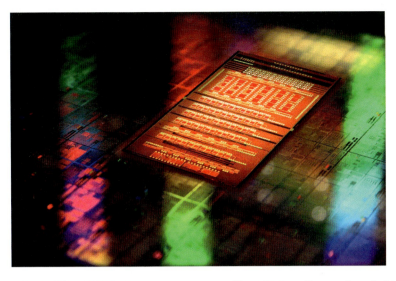

FIGURE 5.67 The SNIPER circuit, presented by IBM on December 1, 2010, which hosts electronic and optical systems on the same piece of silicon, integrated through a communication based on light pulses, rather than on electrical signals

of light. This solution is already used in the transmission of data between different devices, within which electronics still dominates. In the future, integrated circuits will also be of the photonic type and the transition will be characterized by the presence of both components, with the growth of opto-electronic equipment (Fig. 5.67).

Chronology of Astronomers and Physicists and Their Discoveries

Babylonian Astronomers (c. 3000–c. 50 B.C.): Compilation of the *Enuma Anu Enlil* tables (in the days of Anu and Enlil), with the list of the properties of periodic motions of the Sun, Moon, planets and stars.

Wu Xian (c. 1000 B.C.) and **Gan De** (fourth century B.C.): Chinese astronomers who cataloged hundreds of stars.

Thales of Miletus Greek philosopher (fourth century B.C.): Good estimate of the apparent diameter of the Sun and the Moon, each evaluated as the 750th part of its own orbit.

Plato Greek philosopher (428–348 B.C.): Light propagates in a straight line. There are four main colors: black, white, bright, and red.

Aristotle Greek philosopher (384–322 B.C.): Theory of color vision, considered as objective characteristics of bodies. Mixtures of black and white can produce all colors. Model of Earth, believed to be spherical in shape and large.

Aristarchus of Samos Greek astronomer (310–230 B.C.): First heliocentric theory (Earth revolving around the Sun) and estimate of the distances between Earth and the Sun and Earth and the Moon.

Euclid Greek mathematician (third century B.C.): First model of vision based on geometry. The eye, like a lantern, emits visual rays that reach the observed object. The theory sets out simple relationships between the size of objects and the angles formed by visual rays.

Archimedes scientist of Syracuse (c. 287–212 B.C.): Theory of reflection on the basis of the principle of reversibility of the optical path. His treatise (*Catoptrica*), of which there is only indirect information, also contained descriptions of the rainbow and of parabolic (or burning) mirrors, able to concentrate rays into a single point.

Eratosthenes of Cyrene Alexandrian mathematician, astronomer, geographer, and poet (276–194 B.C.): First precise measurement of the size of the terrestrial sphere, based on the position of the Sun.

Hipparchus of Nicaea Greek astronomer, mathematician and geographer (second century B.C.): Models of the motion of Sun and Moon and predictions of solar eclipses; a stellar catalog of 1,080 stars, including their position and brightness, divided in the six groups still used today. Discovery of the precession of the equinoxes.

Hero of Alexandria Alexandrian scientist (first century): Geometry applied to reflection, through a principle of the minimum. Among all the possible rays reflected, the real one is that which minimizes the path. The speed of propagation of the rays is considered infinite.

Claudius Ptolemy Alexandrian astronomer (c. 100–175): Model of the Solar System, centered around Earth (called Ptolemaic or geocentric). First systematic treatment of refraction, with the compilation of a table of refraction angles corresponding to different incidence angles for the pairs water–air, air–glass, and water–glass.

Ibn Sahl Persian mathematician (c. 940–1000): Description of the optics of curved mirrors and of lenses, including the law of refraction, today known as Snell's law.

Ibn Al-Haitham (known as Alhazen), Arab scientist, (965–1038): Founder of modern optics based on the experimental method, with the use of devices such as the camera obscura. In his corpuscular theory, light rays travel towards the eye through small particles emitted by the observed object. In this way, both reflection and refraction are described. He studied the vision process in terms of an optical system—the eye receives the light, the crystalline then focuses the images that the optic nerve transmits to the brain.

Robert Grosseteste English theologian and scientist (1175–1253): Hypotheses about the nature of colors, determined by the intensity of the light beam, from white (the purest), placed before the red, to black, which follows blue. Description of the rainbow regarded as a result of reflection and refraction of sunlight by a "cloud of water."

Roger Bacon English theologian and scientist (c. 1220–c. 1292): Investigation of convex lenses and their use, both for the magnification of small objects and for the correction of seeing defects. The phenomenon of the rainbow is interpreted as a result of the reflection of sunlight by individual raindrops.

Theodoric of Freiberg German theologian and scientist (1250–1310): Explanation of the two arcs (primary and secondary) of the rainbow.

Leonardo da Vinci Italian artist and scientist (1452–1519): Studies of optics and vision.

Nicolaus Copernicus Polish astronomer (1473–1543): Mathematical treatment of the Solar System, with a model centered on the Sun (called Copernican or heliocentric model).

Galileo Galilei Italian physicist and astronomer (1564–1642): Founder of the experimental method and supporter of the heliocentric system; first astronomical observations; and discoverer of four of the moons of Jupiter, of sunspots, and of the stellar nature of the Milky Way.

Hans Lippershey German lens maker (1570–1619), **Charias Jansen**, Dutch spectacle maker (1580–1638) and **Jacob Metius**, Dutch instrument maker (c. 1571–1630): Development of the first microscopes and telescopes, with the objective and eyepiece lenses.

Johannes Kepler German astronomer and mathematician (1571–1630): Model of light that diffuses with infinite velocity and for unlimited distances, with intensity inversely proportional to the square of the distance traveled. Principle of operation of microscopes and telescopes and investigation on the role of the crystalline lens and retina in vision. Formulation of the laws of planetary motion.

Willebrord Snellius Dutch physicist and astronomer (c. 1580–1626): Law of refraction.

Pierre de Fermat French judge and mathematician (1601–1665): Laws of reflection and refraction obtained on the basis of the principle of least time.

Francesco Maria Grimaldi Italian Jesuit priest, astronomer, and physicist (1618–1663): Wave model of light, described as a fluid, on the basis of diffraction experiments through a small opening.

Christiaan Huygens Dutch mathematician, astronomer, and physicist (1629–1695): First wave theory of light, which propagates through an all-pervading substance—the ether.

Robert Hooke English architect and scientist (1635–1703): Model of light as a wave. Improvement of the microscope with the first observations of the anatomy of insects and study of the colors produced by light rays in flakes of mica, soap bubbles, and oil films on water.

Isaac Newton English physicist and mathematician (1643–1727): Corpuscular theory of light, with demonstration that white is a mixture of different colors, which undergo a different refraction when passing through a prism. Study on chromatic aberration in refracting telescopes and the construction of the first reflecting telescope. Formulation of the law of universal gravitation.

Ole Rømer Danish astronomer (1644–1710): First measurement of the speed of light through the observation of the eclipses of Io, one of the moons of Jupiter.

William Herschel British physicist and astronomer of German origin (1738–1822): Discovery of infrared rays, of the planet Uranus, of a large number of double stars, star clusters, and nebulae.

Nicéphore Niépce French inventor (1765–1833): Creator of the earliest form of photograph.

Thomas Young British scientist (1773–1829): Experience of light interference in support of wave theory. First theory of color vision with three types of photoreceptors in the retina, corresponding to red, green, and blue.

Etienne Louis Malus French physicist and mathematician (1775–1812) and **David Brewster**, Scottish physicist and inventor (1781–1868): Investigations into the polarization of light.

Joseph von Fraunhofer German physicist and astronomer (1787–1826): Study of the solar spectrum and diffraction.

Augustin-Jean Fresnel French physicist (1788–1827): Analysis of diffraction and interference of light based on wave theory. Model of light as a transverse wave.

Armand-Hippolyte-Louis Fizeau French physicist (1819–1896): First non-astronomical measure of the speed of the light with an error of only 5 % from the present value. Doppler effect for light waves.

Gustav Kirchhoff German physicist and mathematician (1824–1887): Study of the emission and absorption of light by substances and introduction of the notion of the black body.

James Clerk Maxwell Scottish mathematician and physicist (1831–1879): Theory of electric and magnetic phenomena, and discovery of electromagnetic waves. The speed of propagation of these waves is the same as that of light, which accordingly is identified as an electromagnetic phenomenon.

John William Strutt Rayleigh British physicist (1842–1919): Study on light scattering by microscopic particles, as those suspended in air, and an explanation of the colors of the sky.

Wilhelm Röntgen German physicist (1845–1923): Discoverer of X-rays.

Thomas Edison American inventor and businessman (1847–1931): Development and manufacturing of the electric light bulb.

Albert Michelson (1852–1931) and **Edward Morley** (1838–1923), American physicists: Experimental demonstration of the non-existence of the ether. Accurate measurement of the speed of light.

Max Planck German physicist (1858–1947): Study of the radiation emitted by a black body. Explanation of its spectrum with a revolutionary hypothesis on the oscillators that emit radiation. Their energy does not change with continuity, but as a multiple of an indivisible quantum.

Albert Einstein German physicist, initially a Swiss citizen, then American (1879–1955): Theory of relativity, in which mechanics and electromagnetism present the same symmetry, and where the speed of light has an important role. Hypothesis that light is composed of particles, i.e., photons. Theory of light emission by atoms with the discovery of stimulated emission, the starting point of the laser. Equivalence of mass and energy.

Arthur Eddington British astrophysicist (1882–1944): Experimental verification of the general theory of relativity. Formulation of the stability conditions of a star.

Niels Bohr Danish physicist (1885–1962): Quantum theory of the atom and of its emission, or absorption, of light. Formulation of the wave-particle duality.

Aleksandr Fridman Russian physicist (1888–1925): Theory of the expansion of the universe, today known as the Big Bang.

Edwin Hubble American astrophysicist (1889–1953): Experimental verification of the expansion of the universe.

Theodore Maiman American physicist and electronics engineer (1927–2007): Realization of the first laser.

Arno Penzias American physicist (born in Germany in 1933) and **Robert Wilson** American astronomer (born 1936): Discovery of cosmic background radiation, the most important experimental evidence in support of the theory of the Big Bang.

Photo Credits

Preface
©Reggio Children

Astronomical Computations
©Marsyas/Wikimedia Commons
©Lead Holder/Wikimedia Commons

Alhazen and the Discovery of Light
©Warner Bros. Entertainment Inc.

Kepler's Supernova
©NASA/ESA/JHU/R.Sankrit & W.Blair

The Eclipses of Jupiter's Moon
©NASA

Young and the Wave Nature of Light
©Berit from Redhill, Surrey, UK/Wikimedia Commons

The Röntgen Rays
©Chris Burks and Drondent/Wikimedia Commons
©D-Kuru/Wikimedia Commons
©Old Moonraker/Wikimedia Commons

Microwaves to the Fore
©Ilaria Giuliani

The Light from the Big Bang
©NASA

Waves in Space
©Ralf Roletschek/Wikimedia Commons

Invisible Light
©Scott Nazelrod/Wikimedia Commons
©United States Federal Government/Wikimedia Commons

290   Visible and Invisible

The Speed of Light
©Thomas doerfer/Wikimedia Commons
©United States Federal Government/Wikimedia Commons
©Patrick Stanbro, http://www.patrickstanbro.com
©Judson Brohmer/USAF/Wikimedia Commons
©NASA

Faster than Light
©Patrick McCracken/Wikimedia Commons

The Nature of Light
©United States Federal Government/Wikimedia Commons

Irradiation and Black Body
©Dickbauch/Wikimedia Commons
©Tostaki1/Wikimedia Commons

Reflection and Diffusion
©Oregon's Mt. Hood Territory/Wikimedia Commons

Mirrors
©Deutsche Bundespost

Refraction
©Mehran Moghtadai/Wikimedia Commons

Prisms
©National Portrait Gallery, London/Wikimedia Commons
©Alemas2005/Wikia Lego Message Board

Lenses
©NASA
©Dave Gough/Flickr

Interference
©James Mayer, Arizona State University

Daylight
©SeppVei/Wikimedia Commons
©Suburbia/Wikimedia Commons

Rainbows
©Ajor933/Wikimedia Commons
©CalvinBradshaw, http://photos.calvinbradshaw.com

Photo Credits

Glories
©Franz Kerschbaum, University of Vienna
©Nik Szymanek

Aurorae
©NASA/JPL-Caltech
©NASA
©Senior Airman Joshua Strang, United States Federal Government
©NASA/JSC

The Green Ray
©Mehran Moghtadai/Wikimedia Commons

Lightning
©NOAA Photo Library, NOAA Central Library, OAR/ERL/National Severe Storms Laboratory

Earth's Atmosphere
©Navicore/Wikimedia Commons
©NASA (original image)
©NASA GPM (original image)

The Color of Earth
©Jenny Huang/Wikimedia Commons
©NASA

Space Telescopes, from Hubble Onwards
©NASA

The Sun
©NASA
©JAXA/NASA
©NASA-Goddard Space Flight Center
©Saperaud and Cepheiden/Wikimedia Commons (original image)

Eclipses
©NASA-Goddard Space Flight Center
©Paul Griffin

Stars and Galaxies
©Andrea Dupree (Harvard-Smithsonian CfA), Ronald Gilliland (STScI), NASA and ESA
©NASA, ESA and the Hubble Heritage Team (STScI/AURA)

Why Is the Sky Dark at Night?
©The Hubble Heritage Team (AURA/STScI/NASA)

Light Years
©NASA/JPL-Caltech (original image)

Space Travel
©Donal Davis/NASA
©Carl Sagan, Frank Drake and Linda Salzman Sagan/NASA

Supernovae
©NASA/ESA, The Hubble Key Project Team, The High-Z Supernova Search Team
©NASA, ESA, HEIC and The Hubble Heritage Team (STScI/AURA)

Black Holes
©Dana Berry/NASA
©NASA/CXC/M.Weiss; X-ray: NASA/CXC/CfA/P.Plucinsky et al.; Optical: NASA/STScI/SDSU/J.Orosz et al.

The Active Galactic Nuclei
©HST/NASA/ESA
©NASA and The Hubble Heritage Team (STScI/AURA)

380,000 Years Since the Big Bang
©NASA
©NASA/WMAP Science Team

The Evolution of the Eye
©Opoterser/Wikimedia Commons
©Remember the dot/Wikimedia Commons (original image)

The Human Eye
©Rhcastilhos and Jakov/Wikimedia Commons (original image)

Colour Perception
©Skatebiker/Wikimedia Commons

The Grammar of Colour
©(3ucky(3all /Wikimedia Commons

The Colour Wheel and Harmony
©Piscis13/Wikimedia Commons

Photo Credits

To Deceive the Eye
©Joseph Jastrow/Wikimedia Commons
©John smithson 2007/Wikimedia Commons

Mirages
©Brocken Inaglory/Wikimedia Commons

3-D Images and Movies
©Underwood & Underwood
©Dave Pape/Wikimedia Commons

Birds' Sight
©Dan Pancamo/Wikimedia Commons
©Adrian Pingstone/Wikimedia Commons

Insects' Sight
©Nono64/Wikimedia Commons (original image)
©Dartmouth College

Fish's Sight
©BuzzWoof/Wikimedia Commons
©MiguelCampos/Wikimedia Commons

The Sight of Some Snakes
©H. Krisp/Wikimedia Commons
©Mauro Madonia

Colours and the Survival of the Species
©Honzasoukup/Wikimedia Commons
©PLoS biology
©Adrian Pingstone/Wikimedia Commons

Bioluminescence
©Terry Priest, http://www.frfly.com/Flickr
©Sierra Blakely/Wikimedia Commons
©Dr Steve Miller, United States Federal Government

Photosynthesis
©Danny Mallon, http://www.helpmyphysics.co.uk

The Colours of Leaves
©Jon Sullivan/Wikimedia Commons
©Adrian Pingstone/Wikimedia Commons

The Colours of Other Worlds
©Doug Cummings, Caltech and NASA-Goddard Space Flight Center
©Tim Pyle, Caltech and NASA-Goddard Space Flight Center
©Jon Sullivan/Wikimedia Commons (original image)

Solar Clocks and Sundials
©Edmund Buchner
©François Blateyron, http://www.shadowspro.com

Overhead
©Reggio Children

The Camera Obscura
©DrBob and Pbroks13/Wikimedia Commons
©Sierra Blakely/Wikimedia Commons

Image Manipulation
©Stan Carter
©Rainchill/Wikimedia Commons

Seeing in the Micro World
©Simon Garbutt/Wikimedia Commons (original image)
©Zituba/Wikimedia Commons
©Jan Homann/Wikimedia Commons
©2010 President and Fellows of Harvard College
©Electron Microscopy Facility at The National Cancer Institute at Frederick, United States Federal Government

Photometry
©Hankwang/Wikimedia Commons

The Temperature of Color
©Phrood/Wikimedia Commons (original image)

Thermography
©Mauro Madonia

Diagnostic Imaging
©Carlos Melgoza/Wikimedia Commons
©Mikael Häggström/Wikimedia Commons

Surgery with Images
©Intuitive Surgical, Inc.

LED
©Petey21/Wikimedia Commons
©LumiGram
©Rosa Menkman and Jeroen Joosse/Flickr

Photovoltaic Panels
©US Department of Energy, United States Federal Government
©1997/Wikimedia Commons
©Nick Galante and NASA

Optical Fibers
©Jean-Daniel Colladon/Wikimedia Commons
©Giovanni Piazza

Transmitting Through the Waves
©Toobaz/Wikimedia Commons

Radars
©U.S. Army Space & Missile Defense Command/Army Forces Strategic Command
©Nilfanion/Wikimedia Commons

Satellite Positioning
©yeKcim on the Open Clip Art Library/Wikimedia Commons (original image)
©Savant-fou and NASA /Wikimedia Commons (original image)

Fluorescence and Phosphorescence
©Reggio Children
©FK1954/Wikimedia Commons
©Parent Géry/Wikimedia Commons
©National Institute of Mental Health—National Institute of Neurological Disorder and Stroke, United States Federal Government

Lasers
©Randall L. Ricklefs/McDonald Observatory

Synchrotron Radiation
©Brücke-Osteuropa/Wikimedia Commons
©Michael Krumrey, ESRF

Photonics
©Advanced Micro Devices, Inc. (AMD)/Wikimedia Commons
©IBM Research—Zurich and IBM Archives

Text Quotation Credits

Preface

The experience of Marco and Lucia: Olmes Bisi, Davide Boni, Paola Cagliari, Giovanni Piazza, Maddalena Tedeschi, Vea Vecchi, *Reggio Emilia: children science and creativity Atelier "ray of light", project of the International Center "Loris Malaguzzi"* in *Sociocognitive and Sociocultural Approaches to Science in Early Childhood*, Patakis Publishers, Athens, 2012.

Barbara McClintock in National Research Council, *On Being a Scientist: Responsible Conduct in Research, Second Edition*, pag. 2, Washington, DC: The National Academies Press, 1995.

The hundred languages of children in Carolyn Edwards, Lella Gandini and George Forman Editors: *The hundred languages of children: the Reggio Emilia experience in transformation, 3rd Edition*, pag. 1, Praeger, 2011.

Chapter One: The History of Light

Chapter Opening
Vea Vecchi, Reggio Children

Galileo and the Telescope Pointed Skyward
Galileo Galilei, *Sidereus Nuncius*, translation based on the version by Edward Stafford Carlos, Rivingtons (London 1880), newly edited and corrected by Peter Barker, Byzantium Press, Oklahoma City 2004.

Dante Alighieri, *The Divine Comedy*, Volume 1, Canto II.

Albert Einstein, *Foreword to Dialogue Concerning the Two Chief World Systems* by Galileo Galilei, pag. xvii, University of California Press, Second Revised Edition, 1967.

The Eddington Eclipse
New York Times, November 10, 1919.

Einstein Misunderstood
C. Kirsten and H.-G. Körber (Eds.), *Physiker über Physiker*, pag. 201, Akademie Verlag, Berlin, 1975, quoted in Abraham Pais: *Einstein and the quantum theory*, Reviews of Modern Physics, Vol. 51, No 4 1979, pag. 884.

Anna Auer, *The unknown photographic work of Ferdinand Schmutzer (1870–1928)*. Photoresearcher No 8, September 2005, pag. 17.

R. A. Millikan, Phys. Rev. 7, 355–390 (1916), pag. 384.

Einstein, *Letter to M. Besso*, July 29, 1918; quoted in Abraham Pais: *Einstein and the quantum theory*, Reviews of Modern Physics, Vol. 51, No 4, 1979.

Nobel Foundation Motivation for the Nobel Prize to Albert Einstein.

Abraham Pais, *Einstein and the quantum theory*, Reviews of Modern Physics, Vol. 51, No 4, 1979, pag. 885–886.

The Light from the Big Bang
Ivan Kaminov, quoted in *American Physical Society News* (June 1963: Discovery of the Cosmic Microwave Background).

Chapter Two: Experiments with Light

Chapter Opening
Richard Feynman in R. Feynman, R. Leighton (contributor), E. Hutchings (editor): *"Surely You're Joking, Mr. Feynman!": Adventures of a Curious Character*, W W Norton & Co, 1985.

Electric and Magnetic Fields
Albert Einstein and Leopold Infeld, *The Evolution of Physics : From Early Concepts to Relativity and Quanta*, p. 151, Simon & Schuster, 1966.

James Maxwell, A *Dynamical Theory of the Electromagnetic Field*, Part VI, 1864.

The Nature of Light
Joseph John Thomson, quoted by B. R. Wheaton in *The Tiger and the Shark – empirical roots of wave particle dualism*, Cambridge University Press 1983.

Lenses
Pliny the Elder (23-79 AD), *The Naturalis Historia*, Book XXXVI, Chap. 65 (26) translated by John Bostock, M.D., F.R.S. H.T. Riley, Esq., B.A. London. Taylor and Francis, Red Lion Court, Fleet Street. 1855.

Chapter Three: Light and the Sky

Chapter Opening
Albert Einstein, *Letter to Carl Seelig* (11 March 1952), Einstein Archives 39-013 quoted in *The Ultimate Quotable Einstein*, pag. 20, Princeton University Press, 2010.

Rainbows
Aristotle, *Meteorology*, Book III, Part 2, translated in English by E. Webster.

Glories
Henry Miller, *Big Sur and the Oranges of Hieronymus Bosch*, pag. 95, New Directions Publishing, 1957.

The Green Ray
Jules Verne, *The Green Ray*, Chapt. III, 1882, translated by M. de Hauteville.

Eclipses
NASA Eclipse Web Site (Eye Safety During Solar Eclipses).

Why Is the Sky Dark at Night?
Edgar Allan Poe, *Eureka: A Prose Poem*, Section 7, 1848.

Space Travel
Stephen Hawking, *radio interview with the BBC*, 2006, reported on the online version of the Daily Mail (*"Mankind must colonise other planets to survive, says Hawking"*).

Chapter Four: Light and Life

Chapter Opening
ascribed to Jacques-Yves Cousteau by his son and quoted by Leonard W. Doob in *Pursuing Perfection: People, Groups, and Society*, pag. 65, Greenwood Publishing Group, 1999.

Dalton and the Defects in Colour Perception
John Dalton, *Extraordinary facts relating to the vision of colours*, Memoirs of the Literary and Philosophical Society of Manchester, Vol. V, part 1, pag. 28, 1798.

Mirages
Father Ignazio Angelucci in a *letter to Father Leone Sanzio* (1643), details in Athanasius Kircher, *Ars Magna Lucis et Umbrae* (1646), pag. 801, quoted in Alistair B. Fraser and William H. Mac, *Mirages*, Scientific American, 234, 1, 102–111, 1976.

Bioluminescence
Jules Verne, *Twenty Thousand Leagues under the Seas*, 1871, Part one, Chapter 6 (Full Steam Ahead), translation by William Butcher.
Charles Darwin, *The Voyage of the Beagle*, 1869, Chapter IX.

The Colors of Other Worlds
H.G. Wells, The War of the Worlds, 1898, Book 2, Chapter 2.

Chapter Five: Light Techniques

Chapter Opening
Piero Angela, quoted in Focus n. 111, p. 116, translated by Olmes Bisi.

Image Manipulation
Helmut Gernsheim, *Creative Photography: Aesthetic Trends, 1839-1960*, pag. 244, London: Faber and Faber, 1962, New York: Bonanza Books, 1962, New York: Dover Publications, 1991. Quoted in Michael Busselle, *Complete Book of Photographing People*, pag. 199, Simon & Schuster, 1980.

Glossary

Absorption A phenomenon in which the light beam transfers its energy to the atoms and dissolves.

Big Bang A model that describes the evolution of the universe, constantly expanding and cooling, from about 14 billion years ago; a phenomenon originated by an 'initial singularity,' with very high values of density and temperature.

Black Body An object that absorbs all incident radiation and whose emission spectrum shows a universal trend that depends only on temperature.

Black Hole Celestial body created by the collapse of a star of a very large mass, with gravitational attraction so intense that nothing, not even light, can escape from it.

Coherence Attribute of electromagnetic waves when they are monochromatic and have the same phase and equal polarization.

Color Characteristic of visible light, determined by its wavelength.

Cone Cells Components of the eye, receptors of daylight and color sensitive.

Cosmic Background Radiation Residual of the radiation generated by the Big Bang, detected in the form of microwaves.

Crystalline Lens Natural lens of the eye that focuses light rays onto the retina.

Diffraction A phenomenon in which a light beam is deflected and widened when it encounters an obstacle or an opening of similar size to its wavelength.

Diffusion Light reflection from a rough surface with rebounds in each direction.

Electric Field Property that describes the characteristics of the space around electrical charges.

Electromagnetic Wave Mode of propagation of electric and magnetic fields; by varying the frequency, the electromagnetic spectrum is generated.

Emission Phenomenon in which the energy of the atoms creates a light ray.

Frequency Measure of the number of oscillations of a wave in one second.

Galaxy Huge cluster of stars. The universe consists of approximately 100 billion of galaxies, separated from each other by intergalactic space.

Gamma Rays Electromagnetic waves with a frequency greater than that of X-ray.

Infrared Electromagnetic wave with a frequency greater than that of microwaves and lower than that of visible light.

Interference A phenomenon due to the superposition of two or more waves.

Laser Device that emits a beam of coherent light.

Light-Year Distance covered by light in one year, amounting to 9,461 billion km.

Magnetic Field A property that describes the characteristics of the space around magnets or electric currents.

Microwave Electromagnetic waves with a frequency greater than that of radio waves and lower than that of infrared.

Monochromatic Characteristic of light when it has only one color, i.e., a single wavelength.

Opacity Property of bodies that absorb all electromagnetic radiation.

Photon Constituent of light, a massless particle, endowed with energy, momentum, and spin.

Polarization Property of some electromagnetic waves, called polarized, in which the orientation of the oscillating electric and magnetic fields is constant.

Radio Wave Electromagnetic wave with frequency lower than the microwave.

Reflection Phenomenon in which a light beam bounces off of a smooth surface.

Refraction Event in which a light beam crosses the interface between two different substances.

Retina The inner membrane of the eye formed by different cell types, including rods and cones.

Rod Cells Components of the eye that act as light receptors in poor lighting conditions; insensitive to colors.

Glossary

Spectroscopy Technique of separation of the frequencies that compose an electromagnetic wave and which, in the case of visible light, corresponds to the different colors.

Speed of Light Speed of propagation of visible light and of all electromagnetic radiation, a constant independent from the motion of the observer; insurmountable limit of the speed of bodies and information.

Star Celestial body that emits energy; the Sun is the star nearest to us.

Stimulated Emission An event in which a light beam is emitted under the influence of a radiation of the same frequency.

Supernova Advanced stage in the life of a massive star that explodes with a huge release of energy even 10 billion times that of the Sun.

Transparency Property of the bodies that allow the passage of electromagnetic radiation.

Ultraviolet Ray Electromagnetic wave with a frequency greater than that of visible light and lower than that of X-rays.

Visible Light Electromagnetic wave detectable by the eye, with frequency greater than that of the infrared and lower than that of ultraviolet, corresponding to a wavelength of between 0.7 µm (red) and 0.4 µm (violet).

Wavelength Spatial distance between two maxima of oscillation of a wave.

Wave-Particle Duality Theory on the nature of light, which depending on the phenomenon considered can behave like a wave or a particle.

X-Rays Electromagnetic wave with a frequency greater than that of ultraviolet and lower than that of gamma rays.

Guide to Further Reading

Preface

Carolyn Edwards, Lella Gandini and George Forman Editors: *The hundred languages of children: the Reggio Emilia experience in transformation*, 3rd Edition, Praeger, 2011.

Chapter One: The History of Light

David Falkner, *The Mythology of the Night Sky: An Amateur Astronomer's Guide to the Ancient Greek and Roman Legends*, Springer, 2011.
Phil Simpson, *Guidebook to the Constellations: Telescopic Sights, Tales, and Myths*, Springer 2012.
Galileo Galilei, *Dialogue Concerning the Two Chief World Systems: Ptolemaic and Copernican*, Modern Library, 2001.
Andrea Frova and Mariapiera Marenzana, *Thus Spoke Galileo: The great scientist's ideas and their relevance to the present day*, Oxford University Press, 2011.
Michael Marett-Crosby, *Twenty-Five Astronomical Observations That Changed the World: And How To Make Them Yourself*, Springer, 2013.
Richard J Weiss, *A Brief History Of Light And Those That Lit The Way*, World Scientific Publishing Co., 1996.
Emilio Segrè, *From Falling Bodies to Radio Waves: Classical Physicists and Their Discoveries*, Dover Publications, 2007.
Isaac Newton, *Opticks: Or a Treatise of the Reflections, Refractions, Inflections & Colours of Light*, Dover Publications, 2012.
Abraham Pais, *Subtle Is the Lord: The Science and the Life of Albert Einstein*, Oxford University Press, 2005.

Chapter Two: Experiments with Light

Stan Gibilisco, *Optics Demystified*, McGraw-Hill Professional, 2009.
Ann Breslin and Alex Montwill, *Let There Be Light: The Story of Light from Atoms to Galaxies*, Imperial College Press, 2 edition, 2013.

Richard Dawkins, *The Magic of Reality: How We Know What's Really True*, Free Press, 2012.
James Kakalios, *The Physics of Superheroes*, Gotham, 2006.
Robert Gilmore, *Alice in Quantumland. An Allegory of Quantum Physics*, Springer, 1995.
Richard Feynman, *QED: The Strange Theory of Light and Matter*, Princeton University Press, 2006.
João Magueijo, *Faster Than The Speed Of Light: The Story of a Scientific Speculation*, Cornerstone Digital, 2011.

Chapter Three: Light and the Sky

Bernard Maitte, *Histoire De L'arc-En-ciel*, Editions Du Seuil, 2005 (French).
Peter Grego and David Mannion, *Galileo and 400 Years of Telescopic Astronomy*, Springer, 2010.
David Whitehouse, *The Sun: A Biography*, Wiley, 2005.
Pal Brekke, *Our Explosive Sun: A Visual Feast of Our Source of Light and Life*, Springer 2012.
Lars Lindberg Christensen, Robert A. Fosbury, M. Kornmesser, *Hubble: 15 Years of Discovery*, Springer, 2006.
Stephen Hawking, *The Illustrated A Brief History of Time/The Universe in a Nutshell*, Bantam, 2007.
Brian Greene, *The Elegant Universe*, Vintage Digital, 2011.
Christopher Potter, *You Are Here: A Portable History of the Universe*, Harper Perennial, 2010.
Neil deGrasse Tyson, Donald Goldsmith, *Origins: Fourteen Billion Years of Cosmic Evolution*, W. W. Norton, 2005.

Chapter Four: Light and Life

Betty Edwards, *Color: A Course in Mastering the Art of Mixing Colors*, Tarcher, 2004.
Philip Ball, *Bright Earth: Art and the Invention of Color*, University of Chicago Press, 2003.
Johann Wolfgang von Goethe, *Theory of Colours*, Dover Publications, 2006.
Donald D. Hoffman, *Visual Intelligence: How We Create What We See*, W. W. Norton & Company, 2000.
Simon Ings, *A Natural History of Seeing: The Art and Science of Vision*, W. W. Norton & Company, 2008.
Peter Shaver, *Cosmic Heritage: Evolution from the Big Bang to Conscious Life*, Springer 2011.

Chapter Five: Light Techniques

Stephen Kramer and Dennis Kunkel, *Hidden Worlds: Looking Through a Scientist's Microscope*, HMH Books for Young Readers, 2003.
Spyridon Kitsinelis, *Light Sources: Technologies and Applications*, CRC Press, 2010.
Jane Brox, Brilliant: *The Evolution of Artificial Light*, Mariner Books, 2011.
John Perlin, *From Space to Earth: The Story of Solar Electricity*, Routledge, 1999.
Jeff Hecht, *City of Light: The Story of Fiber Optics*, Oxford University Press, 2004.
Mario Bertolotti, *The History of the Laser*, Taylor & Francis, 2004.

For a Continuous Update

Scientific American, monthly popular science magazine, Nature Publishing Group.

Index

A
Absorption, 124, 200, 240, 243, 261, 287, 301
Achromatopsia, 177
Action at a distance, 57
Active galactic nuclei, 156–159
Additive synthesis, 49, 50
Airy, Sir George, 3
Aldebaran, 139
Alhazen (Ibn Al-Haitham), 11–14, 284
Amplitude modulation (AM), 257–259
Anaglyph, 193
Analogous colors, 182, 183
Andromeda galaxy, 69, 145
Annihilation, 150
Antikythera mechanism, 9–11
Antimatter, 150–151
Apollo 11, 272
Apollo 17, 127
Aqueous humor, 166, 169
Archimedes, 11, 13, 283–284
Aristarchus of Samos, 3–6, 283
Aristotle, 107, 225, 283, 299c
Astraphobia, 122
Astronomical unit (au), 146
Aurora, 113–117, 124

B
Band (Telecommunication), 257, 259
Bentley, Richard, 141
Betelgeuse, 137, 139
Big Bang, 45–47, 69, 140, 146, 160–163, 277, 288, 289c, 292c, 298c, 301, 306
Bioluminescence, 206–213, 293c, 300c
Bird sight, 196–198
Black body, 47, 78–82, 162, 237, 238, 287, 290c, 301
Black body spectrum, 161
Black hole, 68, 153–159, 292c, 301
Blateyron, François, 220, 294c

Blondlot, René, 34–36
Blue hour, 104
Blue marble, 127
Blue supergiant, 139
Bohr, Niels, 75, 287
Bradyon, 70, 71

C
3C 273, 157
Callippic cycle, 11
Camera obscura, 13, 165, 167, 170, 203, 219, 221, 224–227, 284, 294c
Camouflage, 171, 205, 206
Canaletto, 227, 228, 230
Carbon, 151, 210, 211
Carotene, 214
Cassini, Giovanni, 24, 25, 219
Castelli, Benedetto, 15
Cathode rays, 30–32, 34, 279
Chandra X-ray observatory, 20, 155, 157
Chlorophyll, 211, 213, 214
Choroid, 170
Cicero, 11
COBE. *See* Cosmic Background Explorer (COBE) satellite
Coherence, 271, 301
Colladon, Jean-Daniel, 255, 295c
Color, 21–23, 49–53, 61, 84, 88, 98, 101, 102, 104, 109, 116, 124, 133, 137, 163, 166, 171–185, 192, 196, 200, 205, 208, 213, 214, 228, 236, 239, 249, 265, 270, 278, 283, 284, 286, 287, 291c, 294c, 300c, 301–303
Color code, 239
Color constancy, 173
Color temperature, 236–239
Color wheel, 178, 179, 181–185
Columbus, Christopher, 9
Complementarity principle, 75
Complementary colors, 49, 182, 183, 192, 193

Computerized axial tomography, 242, 244
Cone (cells), 169, 171, 175, 196, 200, 301, 302
Contrast (colors), 182, 183. *See also* Simultaneous contrast)
Copernicus, Nicolaus, 285
Cornea, 165, 166, 169, 199
Cosmic Background Explorer (COBE) satellite, 47, 161, 163
Cosmic background radiation, 45–47, 161, 288, 301
Cosmological principle, 140, 163
Crookes tube, 31
Crystalline lens, 165, 166, 168, 170, 198–200, 285, 301
Cyanobacteria, 212, 213

D

Daltonism, 175
Dalton, John, 175–177, 300c
Dante, Alighieri, 15, 297c
Darwin, Charles, 28–30, 206, 209, 300c
Da Vinci, Leonardo, 3, 285
Dawn, 103–105
Democritus, 11
Deuteranopia, 176, 177
Diagnostic imaging, 242–245, 294c
Diffraction, 26, 232, 234, 286, 301
Diffusion, 12, 81, 84, 96, 101, 102, 124, 290c, 301
DNA, 32, 44, 63, 276
Dusk, 103, 104

E

Earth, 3–9, 14–18, 21, 24, 25, 28–30, 38, 54, 61, 64, 65, 101, 103, 113–116, 118, 119, 121, 122, 124, 126, 127, 131, 134, 135, 137, 139, 142–151, 154–156, 158–160, 163, 212, 214, 215, 217, 252, 257, 265, 272, 279, 283, 284, 291c, 306, 307
Earth's atmosphere, 45, 62, 64, 84, 102, 114, 118, 120, 122–124, 212, 291c
Eclipse, 4, 5, 11, 23–25, 37–41, 119, 131, 134–137, 225, 284, 286, 289c, 291c, 298c, 299c
Eddington, Arthur, 37–41, 287, 298c
Èidola, 11–13
Einstein, Albert, 18, 23, 29, 37–39, 41–43, 58, 69, 75, 101, 131, 150, 155, 271, 287, 297c, 298c, 299c, 305

Electric field, 54, 57–59, 301
Electric power, 247, 248, 251
Electromagnetic spectrum, 59–61, 80, 156, 172, 234, 235, 259, 301
Electromagnetic wave, 1, 43, 45, 49, 51, 53, 54, 58–61, 65, 78, 113, 128, 148, 235, 241, 245, 256–258, 260, 262, 265, 275, 280, 287, 301–303
Electron, 31, 32, 41, 63, 72, 74–78, 98, 113–115, 120, 124, 131, 150, 154, 160, 161, 199, 234, 235, 247, 251, 273–275, 278–280, 294c
Electron microscope, 98, 199, 235
Electron volt (eV), 74
Emission, 12, 20, 80, 120, 131, 133, 151, 152, 156, 157, 159, 162, 235, 237, 238, 240, 247, 249, 250, 265, 268, 278, 287, 301–303
Endoscopy, 242
Equinox, 218–220, 284
Eratosthenes of Cyrene, 6–9, 284
Euclid, 3, 12, 283
Evening twilight, 103–105
Event horizon, 155, 156
Evolution, 17, 20, 28, 29, 69, 138, 139, 142, 151, 153, 162, 165–169, 171, 191, 196, 201, 206, 212, 215, 225, 275, 292c, 301
Exoplanet/Extrasolar planet, 128, 148
Exosphere, 124
Experimentum Crucis, 21, 22
Eye, 12–15, 17, 19, 34, 49, 52, 61, 64, 69, 78, 92, 93, 101, 110, 120, 127, 129, 136, 145, 153, 160, 161, 165–173, 176, 178, 179, 185–202, 225, 228–230, 232, 234–236, 246, 248, 260, 283, 284, 292c, 293c, 299c, 301–303

F

Fata Morgana, 190
Fish sight, 200–202
Fluorescence, 31, 211, 241, 265–270, 295c
Focus, 14, 68, 91, 94, 128, 129, 170, 198, 200, 209, 225, 284, 301
Fox fires, 115
Free electron laser, 273
Frequency, 32, 56, 59–61, 63, 74, 80, 81, 111, 120, 167–169, 178, 200, 241, 244, 256–259, 261, 265, 270, 301–303
Frequency modulation, 258, 259

G

Galaxy, 20, 54, 65, 68, 69, 128, 130, 137–140, 143, 145, 146, 148, 149, 152–155, 157–161, 163, 291c, 302, 305
Galilean principle of relativity, 18
Galilei, Galileo, 14–19, 23, 24, 78, 132, 264, 285, 297c, 305, 306
Gamma rays, 59, 63, 64, 74, 84, 121, 124, 129, 156, 160, 302, 303
Garfield, James, 3
General relativity, 37, 40
Geocentric/Ptolemaic system, 15, 18, 19, 284
Geomagnetic storm, 116
Gernsheim, Helmut, 228, 300c
Glass, 30, 32, 33, 52, 74, 87, 89–91, 93, 94, 127, 129, 171, 225, 248, 254, 256, 267, 271, 284
Glory, 109–113, 291c, 299c
Glucose, 210, 211, 214
Gnomon, 1, 2, 217–220
Gravitational lensing, 68
Gravity, 23, 37, 68, 131, 139, 153–156, 158, 265
Green ray, 116–119, 291c, 299c
Grimaldi, Francesco Maria, 26, 286

H

Harmony, 181–185, 292c
Hawking, Stephen, 150, 151, 299c, 306
Heliocentric (Copernican) system, 18, 283, 285
Helium, 29, 131, 138, 151, 160
Hering, Ewald, 175
Herschel, William, 62, 286
Hertzsprung-Russell diagram, 138
HSL diagram, 179
HSV diagram, 179, 180
Hubble deep field image, 128
Hubble, Edwin, 128, 288
Hubble telescope, 20, 65, 127–130, 139–141, 152, 153, 157–159, 291, 292, 306
Hue, 49, 51, 101, 104, 106, 107, 111, 178–181, 183, 196, 214, 228, 236, 238, 239
Huygens, Christiaan, 25, 26, 286
Hyades (star cluster), 40
Hydrogen, 29, 131, 138, 139, 151, 160, 211

I

Illumination, 52, 98, 208, 235, 236, 248
Image manipulation, 228–229, 294c, 300c
Infrared, 20, 21, 44, 45, 51, 59, 61, 62, 74, 78–81, 128, 129, 156, 161, 162, 200, 202–204, 216, 239–241, 249, 286, 302, 303
Infrared camera, 239–241
Insect sight, 198–200
Integrated circuit, 279, 281
Interference, 26, 27, 46, 75–78, 95–99, 111, 286, 290c, 302
Interference grating, 95, 96
Io (satellite of Jupiter), 23–25, 286
IOK-1, 145
Ionization, 32, 44, 63, 64, 120, 124, 244, 245, 276
Ionosphere, 115, 124, 257
Iris, 166, 169–171
Irradiance, 235

J

Jastrow, Joseph, 186, 293c
Jupiter, 14, 15, 23–25, 128, 285, 286

K

Kelvin temperature, 161, 162, 236
Kepler, Johannes, 14, 19, 20, 153, 285
Kirchhoff, Gustav, 81, 82, 287

L

Laser, 43, 270–273, 287c, 288c, 295, 302
LEDs, 247–251, 295c
Lens, 13, 49, 55, 68, 91–94, 170, 197, 221, 225, 230, 232, 277, 284, 285c, 301
Light meter, 236, 237
Lightness, 178, 179
Lightning, 120–122, 291c
Lightning conductor, 121
Light-year, 20, 21, 65, 69, 140, 144–146, 155, 157–159, 292c, 302
Local group, 145
Loys de Cheseaux, Jean-Philippe, 142
Luciferase, 208
Luciferin, 208
Lumen, 236, 248–250
Luminous power, 236, 248, 249, 251, 252
Luxon, 70, 72

M

M84, 158
Magellanic clouds, 145, 153
Magnetic field, 53, 54, 56–59, 113–115, 124, 132, 245, 298c, 301, 302
Magnetosphere, 114, 116, 124
Magnetron, 43, 44, 59
Malaguzzi, Loris, 297c
Marconi, Guglielmo, 257
Mather, John, 47, 163
Maunder minimum, 134
Maxwell equations, 58
Maxwell, James Clerk, 58, 59, 287, 298c
McClintock, Barbara, 297c
Mercury, 37
Mesosphere, 123
Messier 100 galaxy, 130
Meton's cycle, 9
Microscope, 93, 98, 199, 230, 232–235, 267, 270, 285, 286
Microwave, 43–47, 59, 150, 162, 163, 256, 260, 273, 289c, 298c, 301, 302
Microwave oven, 43–45, 59
Milky Way, 20, 128, 129, 140, 145–148, 153, 159, 161, 285
Miller, Henry, 112, 299c
Millikan, Robert, 41, 78, 298c
Mirage, 188–191, 293c, 300c
Mirror, 6, 13, 52, 84–86, 96, 192, 221, 225, 272, 284, 290c
Monochromatic, 22, 97, 177, 200, 270, 271, 301, 302
Moon, 3–6, 9, 11, 15, 16, 24, 38, 65, 73, 91, 107, 134, 135, 146, 147, 149, 236, 238, 272, 277, 283, 284
Moon rainbow, 107, 108
Moore's law, 279
Morning twilight, 103–105
MRK 205, 159
M33 X-7, 155–157

N

N63A, 153
Nanoscience and Nanotechnology, 33, 254, 273, 275, 277
Natural background ionizing radiation, 64
Natural irradiation (Electromagnetic), 257
Neurons, 175
Neutron star, 154
Newton, Isaac, 21–23, 26, 28, 37, 38, 78, 90, 134, 141, 142, 181, 286
Newton's corpuscles, 23, 26
NGC 1300, 140
NGC 4261, 158
NGC 4319, 159
NGC 4526, 152
Nimrud lens, 94
N-rays, 34–37
Nuclear magnetic resonance, 245
Nuclear reactions, 78, 131, 150

O

Objective, 127, 232, 285
Ocelli, 165, 166
Ocular, 232
Olbers' paradox, 142, 143, 223
Olbers, Wilhelm, 142
Ommatidia, 198, 199
Opacity, 124, 302
Opponent process, 172, 173, 175
Optical fiber, 242, 254–256, 280, 295c
Optic nerve, 14, 171, 198, 199, 284
Opto-electronics, 281
Overhead projector, 221–224
Oxygen, 93, 123, 151, 208, 210–213
Oxygen catastrophe, 212

P

Pais, Abraham, 43, 298c
Parallax, 195
Parsec, 146
Penzias, Arno, 45–47, 162, 288
Perigal, Henry, 2
Period, 11, 56
Perspective, 191, 225
Phosphorescence, 208, 241, 250, 265–270, 295c
Photometric quantities, 235, 236, 248
Photometry, 235–236, 294c
Photon, 23, 41–43, 63, 66, 70, 72–77, 160, 161, 208, 265, 274, 280, 287, 302
Photonics, 279–281, 295c
Photophores, 200, 208
Photosynthesis, 210–215, 293c
Photovoltaic panel, 236, 251–254, 277, 295c
Pioneer 10, 149
Pixel, 228, 230
Planck, Max, 41, 287
PLANCK satellite, 163
Planck's constant, 74
Plato, 12, 283
Pliny the Elder, 94, 299c

Index 313

Poe, Edgar Allan, 142–144, 299c
Pointillism, 184, 185
Polarization, 55, 165, 286, 301, 302
Posidonius, 8, 9
Positron, 75, 151
Positron Emission Tomography (PET), 151
Primary colors, 23, 49–51, 177, 181, 182, 228
Prism, 21, 22, 34, 36, 49, 62, 78, 88–91, 106, 134, 286, 290c
Probability wave, 75
Protanopia, 176, 177
Proxima Centauri, 146, 149
Ptolemy, Claudius, 9, 284
Pupil, 12, 127, 170
Pythagoras' theorem, 1–2

Q
Quantum well, 250
Quasar, 40, 157, 159

R
Radar, 43, 46, 259–261, 295c
Radarange, 44
Radiography, 31, 33, 64, 120, 242–244, 277
Radiometric quantities, 236, 251
Radio wave, 34, 40, 45, 59, 69, 74, 84, 124, 129, 156, 157, 242, 244, 257–260, 262, 302
Rainbow, 21, 23, 89, 95, 98, 99, 104–109, 176, 181, 284, 285, 290c, 299c
Red giant, 139, 151
Red supergiant, 138, 139
Reflection, 18, 22, 55, 84, 85, 87, 96, 107, 109, 256, 283–285, 290c, 302
Reflex apparatus, 225, 226
Refraction, 22, 87–89, 91, 106, 107, 109, 119, 188, 190, 254, 284–286, 290c, 302
Retina, 136, 170, 171, 175, 184, 285, 286, 301, 302
RGB model, 49, 182, 228, 230
Rigel, 139
Rod cells, 302
Roddenberry, Gene, 151
Rohmer, Eric, 119
Rømer, Ole, 24, 25, 286
Röntgen, Wilhelm, 30–34, 279, 287

S
Saros cycle, 11, 135
Satellite positioning, 262–265, 295c
Saturation, 178–180
Saturn, 14, 146
Scheiner, Christoph, 18
Sclera, 171
Secondary colors, 49–51, 182
Serpentarius (constellation), 19
Seurat, Georges, 184, 278
Sexual selection, 206
Signac, Paul, 184, 185
Silicon, 93, 151, 237, 250, 254, 279, 281
Simultaneous contrast, 173–175
Sky, 1–3, 14, 16, 19, 39, 55, 101–104, 107, 119, 124, 128, 137, 141–144, 163, 165, 189, 190, 239, 287, 292c, 299c
Smoot, George, 47, 163
SN 1604, 19, 20
Snake infrared sight, 202–204, 293c
SN 1994D, 152
Snellius, Willebrord, 8, 285
SOHO satellite, 114
Solar cell, 251, 252
Solar chromosphere, 132, 133
Solar clocks, 217–221, 294c
Solar core, 132
Solar corona, 131, 132, 136
Solar energy, 131, 212, 252, 253
Solar flare, 116, 132
Solar photosphere, 131, 132
Solar prominence, 132
Solar wind, 113–116, 124, 131
Solstice, 7, 8, 217, 219, 220
Space shuttle, 73, 129
Spectral colors, 101, 106, 107
Spectroscopy, 303
Speed of light, 25, 59–61, 65–72, 75, 120, 143, 145, 149, 151, 260, 274, 277, 280, 286, 287, 290c, 303
Spencer, Percy, 43
Spitzer Space Telescope, 20
Star, 1, 14–16, 19–21, 24, 28, 36–40, 54, 68, 69, 74, 104, 123, 128, 131, 134, 135, 137–144, 146, 148–160, 215–217, 236, 273, 283, 284, 286, 287, 291c, 301–303
Stereoscopy, 191, 192, 246
Stimulated emission, 43, 271, 287, 303
Strabismus, 196
Stradivari, 134
Stratosphere, 123
Subtractive synthesis, 49–51, 182, 184

Sun, 1–3, 5–9, 11, 14–16, 18, 20, 21, 25, 28–30, 36–40, 53, 54, 61, 62, 65, 73, 78–81, 94, 101, 103–105, 107, 109, 110, 112–116, 118–120, 123, 124, 126, 127, 131–140, 142, 146–149, 152, 154–158, 165, 189, 200, 215–221, 236, 251, 252, 280, 283–285, 291c, 303
Sundials, 217–221, 294c
Sunspots, 14, 16, 18, 132, 134, 285
Superman's X-ray vision, 12, 120
Supermassive black hole, 158, 159
Supernova, 19–21, 121, 128, 139, 151–154, 156, 292c, 303
Surface wave, 110, 111
Synchrotron radiation, 33, 273–278, 295c

T
Tachyon, 70, 71
Terrestrial gamma rays, 121
Tertiary colors, 181, 182
Thermal agitation, 80
Thermal equilibrium, 240
Thermography, 239–241, 294c
Thermosphere, 124
Thomson, Joseph, 32, 76
Thomson, William (Lord Kelvin), 28, 29
Thunder, 120–122
Total reflection, 254, 255
Transistor, 279, 280
Transparency, 82–84, 124, 126, 242, 303
Triangulum Galaxy, 145
Tritanopia, 177
Troposphere, 122

U
Ultrasonography, 244
Ultrasound, 244
Ultraviolet ray, 34, 49, 51, 59, 79, 81, 84, 114, 123, 124, 128, 131, 156, 166, 169, 196, 197, 200, 234, 241, 250, 265–267, 269, 303
UVC, 61, 63

V
Value, 178–180
Varying speed of light theories, 69
Veduta, 225

Venus, 14
Verne, Jules, 119, 208–210, 299c, 300c
Video surgery, 242, 245–247
Virgo supercluster, 145
Visible light, 1, 20, 21, 59, 63, 69, 79, 81, 84, 95, 120, 124, 128, 129, 133, 156, 162, 203, 204, 234–236, 238, 239, 241, 242, 259, 260, 265, 267, 268, 289c, 301–303
Vitreous humor, 165, 168
von Goethe, Wolfgang, 181, 306c
Voyager 2 spacecraft, 24

W
Wave, 26, 27, 34, 46, 51–56, 58–60, 72, 73, 75–77, 88, 95–98, 110, 111, 120, 121, 232, 244, 256, 259–261, 270, 271, 286, 287, 289c, 295c, 299c, 301, 303
Wavelength, 33, 44, 49–52, 54, 56, 59–61, 74, 80, 84, 87, 95, 97, 98, 111, 124, 133, 134, 144, 161, 171, 172, 178, 213, 232, 234, 235, 238, 242, 249, 256, 259, 273, 301–303
Wave-particle duality, 76, 288, 299c, 303
Wells, Herbert, 216, 300c
Wheatstone, Charles, 191, 192
White dwarf, 139, 151
White light, 21–23, 49, 53, 80, 89, 98, 101, 104, 106, 118, 171, 249
Wilson, Robert, 45–47, 162, 288
WMAP satellite, 163, 292c
Wood, Robert, 34, 36, 37
Wood's lamp, 63, 266

X
X-Rays, 10, 12, 20, 21, 30–34, 61, 63–65, 84, 120, 124, 133, 155–157, 234, 242–244, 273, 275, 277, 287, 292c, 302, 303

Y
Young's interference, 26, 27
Young, Thomas, 26, 286

Z
Zoophyte, 209

Printed by Printforce, the Netherlands

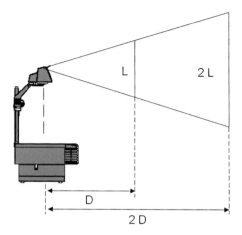

FIGURE 5.10 If the distance (D) between the OHP and the screen doubles (2D), the linear dimensions of the projected image (L) will vary in the same way (2L)

> **CONSIDER THIS**
> The confectionery industry uses machines for spraying various ingredients: sugar, chocolate, jam, flavorings, etc. The amount of substance spread over a surface depends how far away the sprayer is and follows the same law as light rays. Can that be true?

The Camera Obscura

Light rays are diffused from each illuminated point and spread in all directions along straight paths. What happens if the light given off by an object, for example, the tree in Fig. 5.11, heads to a box with an opening on one of its sides?

Most of the luminous rays are absorbed or reflected by the wall of the box, but some of them travel in the right direction to enter the hole. If the inside of the box is dark, on the wall opposite the opening an image is formed. By examining some of these trajectories you will realize that the picture on the wall is... upside down!